MYSTERY ANIMALS OF THE BRITISH ISLES

Gloucestershire and Worcestershire

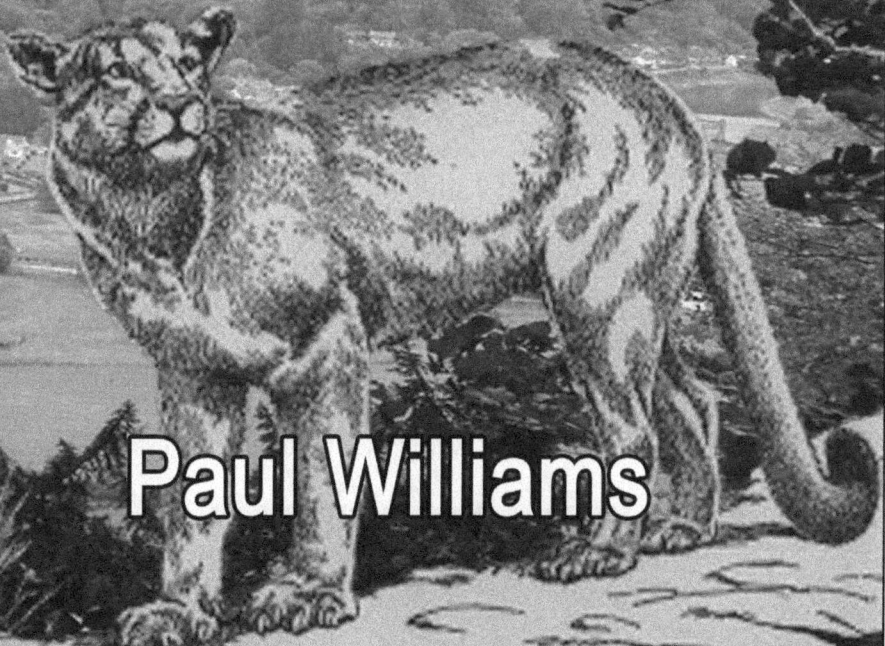

Paul Williams

Typeset by Jonathan Downes,
Cover and Layout by SPiderKaT for CFZ Communications
Using Microsoft Word 2000, Microsoft , Publisher 2000, Adobe Photoshop CS.

First published in Great Britain by CFZ Press

CFZ Press
Myrtle Cottage
Woolsery
Bideford
North Devon
EX39 5QR

ISBN: 978-1-905723-73-7

CONTENTS

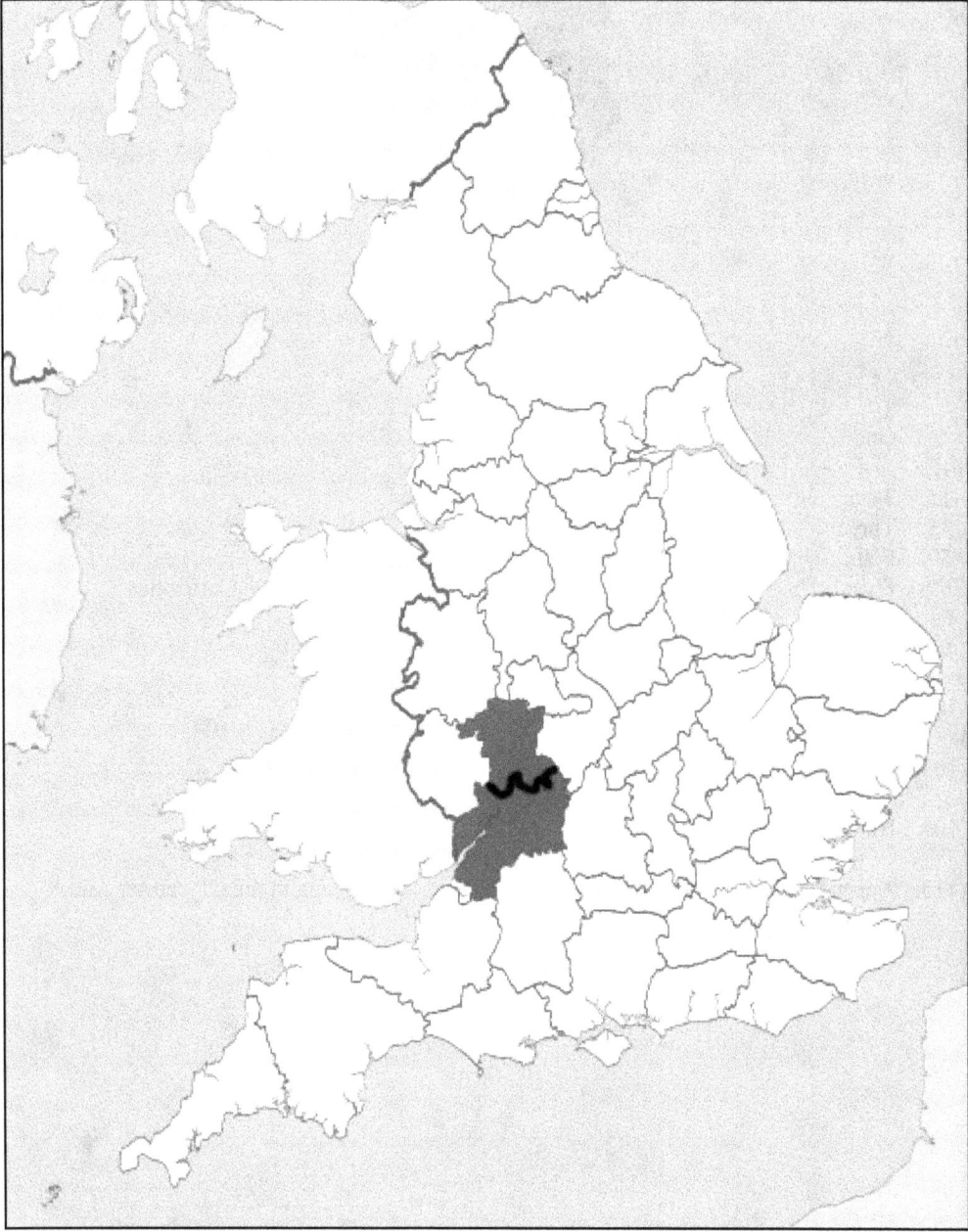

ONE

Many Mysteries

At the age of eight I discovered several books about the yeti in my local library and read them all, returning to borrow a selection for a second and, sometimes third, perusal. Scientists such as Bernard Heuvelmans and Ivan Sanderson became my heroes and role models, inspiring a continuing personal fascination with the search for the unknown, which in later years I have realised is actually both a search for truth and a desire to challenge accepted beliefs that are not based on facts. My interest in yeti literature resulted in an optimistic proposal in 1996, twelve years after the first library forage, that my undergraduate dissertation for a history degree should be about the search for mysterious hominids. My tutor at the University of Teesside surprisingly declared this a suitable topic and awarded an A grade to the completed essay which had expanded beyond the Himalayas to discuss the North American Bigfoot, the Mongolian alma and various other similar creatures that had been, and continue to be, sighted around the world.

My faith in the educational system temporarily restored I progressed to complete a Masters degree in Mediaeval Studies at a more traditional university where the dissertation's topic was regarded with scepticism if not derision. Part of my academic studies covered the Vikings and their battles with the forces led by King Alfred the Great. At that time there was a huge controversy over the publication of a book by Alfred Smyth entitled *The Mediaeval Life of King Alfred the Great*. Smyth suggested that the biography of King Alfred supposedly written by a tenth-century monk named Asser was a later fake. Some academics who criticised this hypothesis did so not on technical grounds but from a preconceived notion that it had to be wrong because it was contrary to the established belief. In my mind I was convinced that all arguments should solely be based on evidence.

In 1999 Nick Pope wrote a book about UFOs called *Open Skies, Closed Minds.* I felt that some members of the traditional academic community had closed minds when it came to theories that challenged existing long-held beliefs. Instead of searching for new ideas they seemed content to accept and defend the current dogmas. As it happens I was not convinced by the evidence put forward by either Smyth or Pope but read both books before reaching those conclusions. There are countless other examples, in many areas, where people and institutions that represent established views automatically block, or attempt to block, notions that challenge those views. This is one of the reasons why Cryptozoology still has no official recognition.

The University of Sheffield did allow me to compile a doctoral thesis entitled *Cultural Connotations of the Wolf in England* which built on the work of professional zoologists in challenging the widely held view that wolves are habitual man-eaters. The thesis became the basis for my first book, *Howls of Imagination*, which was published in 2007. By then I had, for several years, been a member of the Centre of Fortean Zoology, which is the only full-time professional scientific organization dedicated to Cryptozoology in the world. When I discovered that they were compiling a series of books on Britain's mysterious animals I put my name forward for the volumes covering Gloucestershire and Worcestershire.

These two counties roughly equate to the Mercian Kingdom of Hwicce, created in 628, but my reasons for combining them again are personal rather than historic. After getting married in the autumn of 2006 I came to live, literally, on the border of Gloucestershire and Worcestershire, rejecting urban living for a more rural setting. Like much of Britain's countryside my neighbourhood is totally devoid of street lamps and it is possible when walking along a deserted road at night to convince yourself that all manner of creatures are observing your every movement. This provides rare moments of solitude to temporarily empathise with our ancestors who were rarely able to illuminate the night and had good reason to be wary of dangerous predators lurking in the blackness.

Gloucestershire lies at the top of South West England. It borders Herefordshire, Worcestershire, Warwickshire, Oxfordshire, Somerset, Wiltshire and Avon. The town of Gloucester, the county's capital, was founded by the Romans in AD48. William the Conqueror visited in 1085 and, whilst there, ordered that the Doomsday Book be written. Gloucester was granted its first charter in 1155 by King Henry II and gradually developed into a busy town and port. For a while the larger port of Bristol was also included in Gloucestershire's borders, before being incorporated into Avon although Gloucestershire's county cricket team still has its headquarters in Bristol. Apart from Gloucester the principal towns in Gloucestershire are Cheltenham, Stroud, Cirencester and Tewkesbury.

Worcestershire is on the edge of the West Midlands. As well as Gloucestershire it borders Herefordshire, Shropshire, Staffordshire and Warwickshire. Worcester is the capital city, founded by the Romans around the same time as Gloucester. In 680 AD Worcester was given a bishop and cathedral where King John was buried in 1216. During the English Civil War the cathedral was used as a stable by the parliamentary army which occupied Worcester. Other towns in Worcestershire include Bromsgrove, Evesham, Kidderminster, Malvern and Redditch.

Between them Gloucestershire and Worcestershire contain roughly one percent of Britain's population and two percent of its total area. I have attempted to keep this book with the confines of that area but occasionally stray across the borders. Both counties are predominantly rural but sightings of mysterious animals are not confined to the countryside.

The title of the book is something of a misnomer as there is nothing mysterious about many of the animals and birds referred to. The quarry sought by Cryptozoologists can be divided into two categories, animals which are unidentified and those which are incongruous. Most of those referred to in this book belong in the second category. The mystery surrounds their presence in Gloucestershire and Worcestershire, whether verified or merely reported. Perhaps hidden animals would be a better title since most of the creatures described, if real, are certainly adept at hiding from man.

To many people the idea that any unidentified or incongruous animals could reside in the British Isles, even in the less populated counties, is preposterous. And yet there are confirmed instances of rare and unusual creatures being found in Gloucestershire and Worcestershire plus a host of anecdotal reports, often from credible witnesses. Amongst the verified creatures are beavers, various insects, wild boar, a funnel web spider, a porpoise, terrapins and tortoises. Those seen include panthers, pumas, lynx and a crocodile.

Sometimes the animals have escaped from private collections. This was the case with the African porcupine, Spike, who was captured at a bus stop at Knockly patch, Bream in the Forest of Dean on 9 January 2005.[1] During a storm he had escaped from a farm in Elwood, near Coleford. He had been staying temporarily on the farm with his family. The owner of the farm, Liz Black, rounded up two of the other porcupines but could not prevent Spike's getaway. The authorities did not return him to Ms Black because she did not have a licence for wild animals. She claimed not to know that one was needed, a belief that was three decades out of date.

Until 1976 it was entirely legal to keep dangerous wild animals in Britain without a licence. People who had travelled around the British Empire, and beyond, brought back live souvenirs or those who were rich imported animals such as peacocks, zebras and big cats. In 'The Adventure of the Speckled Band', Sherlock Holmes encountered the evil Doctor Grimesby Roylott who kept a cheetah[**], baboon and a deadly snake in his country home. This encounter, set in 1883, was invented by Sir Arthur Conan Doyle but illustrates the tendency of some people to keep exotic creatures.

The throwback to colonial days was ended with the passing of the Dangerous Wild Animals Act in 1976 which mandated not only a license but also adequate insurance for certain animals considered to be dangerous. This was modified in 1984 to include hybrids. Some researchers believe that the big cats now seen across the country are the descendants of animals released by landowners unwilling to pay for a licence or who felt that they would not meet the criteria

---

[**] The cheetah was replaced by a leopard in the classic 1984 adaptation starring Jeremy Brett. Incidentally the 'swamp adder' kept by Dr. Roylott was a made-up beast with certain similarities to various kinds of venomous snake including pit vipers and cobras.

specified by the act. This is perfectly plausible, especially given the requirements for the premises to be inspected and the license renewed annually. Some people have since admitted that they did release animals into the wild in the 1970s. Every year hundreds of pet dogs and cats are taken from their homes and dumped by uncaring owners. In mediaeval Europe parents who could not afford to look after their children, or were embarrassed by their illegitimacy, or disability, would sometimes leave them to die in the woods. Many fairy tales touch on this theme and the practice regrettably continues in some countries today.

In keeping with most recent British legislation the Dangerous Wild Animals Act contained a naïve loophole. This was not rectified for a further five years when the Wildlife and Country-side Act made it illegal to release certain animals into the wild. In keeping with most recent British legislation this is often ignored. Illegally imported animals seized by British Customs Officers in 2007 included an alligator, turtles, tortoises and a tiger cat.[2] There were three known releases of non-native species in Britain in 2007, nine in 2008 and seven in the first seven months of 2009.[3] These are just the ones that were discovered. A charity in Essex, the Dangerous Wild Animal Rescue Facility, receives around twenty calls a day and has housed animals such as Scottish wildcats and alligators. On 4 December 2009 the Daily Mail reported that officials working for the UK Border Agency, formerly Customs, had found three alligator lizards hidden inside video cassettes. A spokesman said that sending live animals by courier and through the post was becoming more common. Perhaps this is because Britain has reduced the number of Customs Officers to a bare minimum and removed a permanent presence entirely from some ports and airports in favour of what is described as an intelligence led approach.

Some conscientious citizens do comply with the law by obtaining licenses to keep wild animals, with prices and conditions disgracefully varying between local authorities. Two examples come from 2000 when South Gloucestershire Council issued a licence for someone to keep three wolf dogs and Cotswold District Council issued a similar licence to a man in Cirencester who said that he intended to breed pure wolves.[4] This claim was doubted because the local animal warden believed that his pure female timber wolf was from a husky breed. Nevertheless the man claimed to have a contract to supply wolves to Sweden for £2000 and charged £600 per pup in Britain. There were rumours in South Gloucestershire that the licensed wolf dogs had bred and that there were other unlicensed animals in the area.[5] Guidance on the act can be read on the websites for most local authorities.

On 6 November 2002 Mr Norman Baker MP asked the Secretary of State for Environment, Food and Rural Affairs, the following questions:

> 1. How many primates are owned under a Dangerous Wild Animals Act 1976 licence and for what purpose these animals are owned?
>
> 2. How many animals are owned under a Dangerous Wild Animals Act 1976 licence, broken down by species?
>
> 3. If she plans to amend the 1984 schedule to the Dangerous Wild Animals

Act 1976 to remove animals that have been shown to pose no serious danger to the public?

The reply came from Mr Elliott Morley, the Parliamentary under-secretary for DEFRA (now famous for his expenses claims) as follows:

As local authorities administer the Dangerous Wild Animals Act 1976 details of licensed animals are not held centrally. However, as part of a review to examine the effectiveness of the Act, consultants reported that on the base of a 95 percent response from local authorities, 655 primates were then licensed. They did not seek the reason for keeping these animals.

The consultants found that a total of 11,878 animals were licensed, categorised as follows;

Category	Number
Primates	655
Rodents	14
Carnivores	269
Pinnipeds	3
Elephants	1
Odd-toed Ungulates	20
Even-toed Ungulates	5339
Birds	5232
Reptiles	334
Invertebrates	11

The Even-toed ungulates and birds comprise for the most part, farmed wild boar and ostrich. A copy of the consultants report can be downloaded by visiting http://www.defra.gov.uk/wildlife-countryside/consult/dwaa/index.htm.[6]

The consultants' report has recommended changes to the 1984 schedule to the Act. We have sought views from the public on all their recommendations, including those on the schedule and shall be drawing up our own proposals to address the shortcomings identified. Our proposals will be the subject of a further public consultation exercise in due course, as a result of which decisions will be taken as to whether or not the schedule needs to be altered."[7]

Mr Morley's reply relates to the known creatures. A glimpse of the unknown comes from a survey commissioned by Disney to mark the release of their film *The Wild* in 2006. This cartoon is about some zoo animals that, as the title suggests, go to live in the wild. The survey was compiled by animal and conservation groups, including Beastwatch UK which records sightings of unusual creatures around Britain. In total the survey noted the following sightings:

5931 Big cats (panthers, pumas, leopards, lynx and jungle cats),
3389 Sharks,
332 wild boars,
51 wallabies,
43 snakes,
15 owls,
13 dangerous spiders (including a tarantula and black widow),
13 raccoons,
10 crocodiles,
7 wolves,
4 eagles,
3 pandas,
2 scorpions
1 penguin.[8]

Between 2001 and 2007 there were one hundred and twenty five reported sightings of exotic animals, involving one hundred and sixty five individual animals, to the Rural Development Service and, its successor, Natural England.[9] Of these, twenty five were confirmed, thirteen were supported but not confirmed, thirteen were discounted, eleven were attributed to other causes and sixty three were inconclusive.

Only 19%, or one in five were dismissed.

The full list is as follows:

Coypu:

Eleven reported cases.
Twelve animals.
Four cases discounted.
Three cases attributed to other causes.
Four cases inconclusive.

Wild Boar:

Thirty one reported cases.
Sixty Six animals, until January 2007 when the presence of boar was accepted in some areas.
Eighteen cases confirmed.
Eight cases supported but not confirmed.
Two cases discounted.
One case attributed to other causes.
One case inconclusive.

Big Cats:

Sixty reported cases.
Sixty animals.
Two cases supported but not confirmed.
Seven cases discounted.
Five cases attributed to other causes.
Forty six cases inconclusive.
Other Exotic Cats:
Two reported cases.
Two individual animals.
Two cases inconclusive.

Siberian Chipmunk:

Six reported cases.
Eight animals.
Five cases confirmed.
One case inconclusive.

Other:

Fifteen cases.
Seventeen animals.
Two cases confirmed.
Three cases supported by evidence but not confirmed.
Two cases attributed to other causes.
Eight inconclusive.

The above data either makes a mockery of the commonly held belief that all the animals, in the United Kingdom are known and documented or suggests that there are thousands of people with poor judgments and eyesight.

This book looks, briefly, at a range of incongruous animals, birds and insects found or reported in Gloucestershire and Worcestershire.

Map 1: Gloucestershire and Surrounding Area

Map 2: Worcestershire

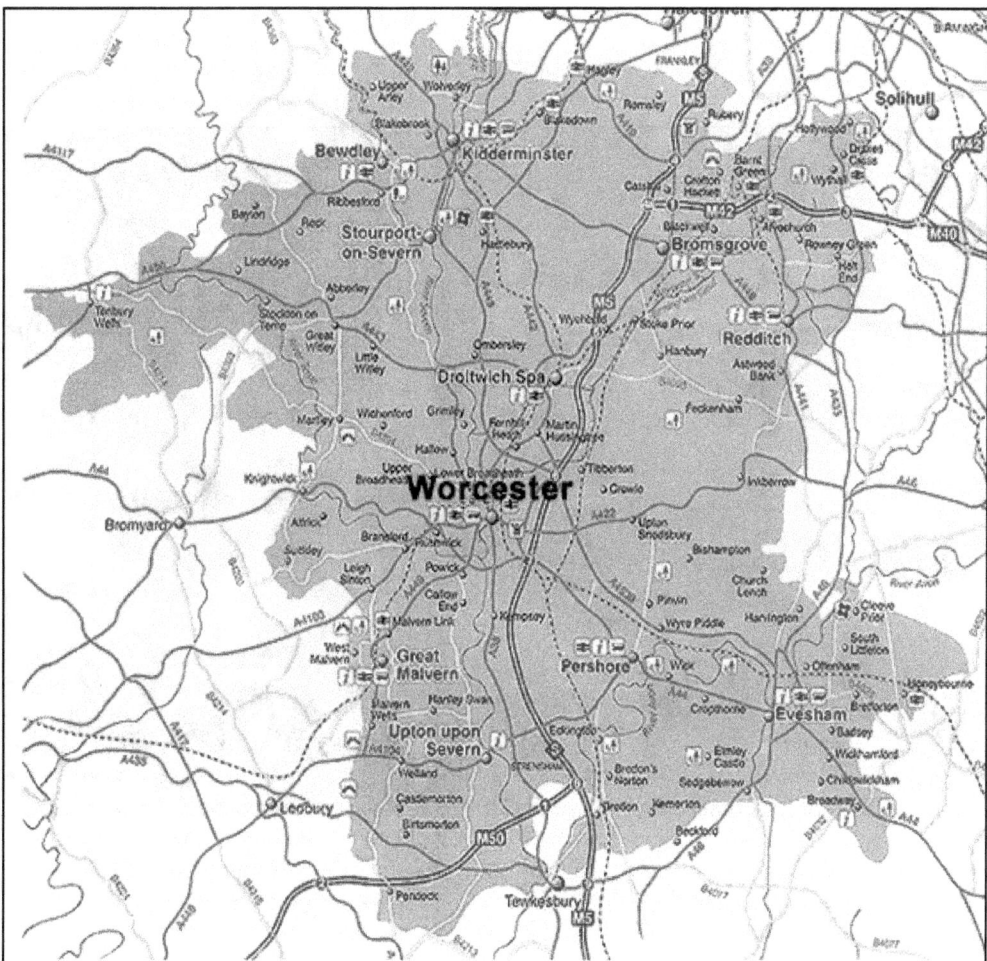

A Lion in Tanzania but could there really be one in Ashchurch near Tewkesbury? Photo by Richard Spencer, 03 May 2008, image in the public domain downloaded from http://www.publicdomainpictures.net/view-image.php?image=575&picture=resting-lioness, 17 May 2010.

Two
Fantastic Felines

The most commonly sighted incongruous creature is the big cat. Officially Britain is home to many species of domesticated cat, and a few feral ones, but not any larger species. The lynx was eradicated around five centuries ago, the date of the extinction was originally considered to be three and a half thousand years earlier, although there are serious suggestions that they should now be reintroduced to cull the expanding deer population. In Scotland the same reason has been proposed to reintroduce wolves. Other European countries, in line with the Bern Convention and the Habitats Directive, have seriously considered this. France, Italy and Austria are amongst those who have tried to reintroduce the Lynx.

Conferences examining the potential and pitfalls of reintroducing wild animals in Britain have been held, including one in Cirencester in 2006. However, anyone expecting progress seems doomed to disappointment. The vagaries of the British political system, the unwanted and costly influences of unelected quangos, the slowness of the legislative process and the persistent official reluctance to sanction anything that does not contain an immediate potential for profit, will surely condemn the return of large predators to the pages of wishful thinking.

Unless of course the lynx, and other predators, are already here. Sightings of incongruous cats in the countryside and occasionally, the towns, predate the passing of the Dangerous Wild Animals Act. In 1962 a huge cat that killed domestic felines and lambs was shot at Great Witley in Worcestershire. It was described later, by a witness as twice as big as a large tomcat with a short blunt ended tail and long fangs that stuck out from the side of its mouth.[10] The witness was fourteen at the time and went on to become an unofficial vermin killer for local farmers. He had not seen anything like this cat before or since.

In 1977 a Mrs Armitage was driving between Eastnor and Upton upon Severn, along a road close to my present home, when she saw a puma. She told researcher Marcus Matthews, albeit long after the event, that the animal was dark with a small head and long tail. It was higher at the shoulder than a lioness.[11]

On 2 August 1986 the *Daily Telegraph* reported an encounter between motorcyclist, David Hart, and a big cat which was the size of a large dog with a cat's head and long tail. In two letters written to Marcus Matthews in 1988 Mrs Louise Cook who lived two miles from Symonds Yat near Ross-on-Wye described an encounter with a wild cat.[12] The incident happened on her farm at around 15:30 on a warm summer's afternoon. She believed that the animal was a Scottish wild cat and reported the incident to her vet and a man named Mike Tomkins who lived in Scotland and studied the wild cat. One of her friends had said that there were wild cats in nearby woods.

On 7 November 1993 the *Sunday Times* reported that an animal described as a jungle cat was seen six times in and around Kidderminster in 1992. Douglas Richardson, the animal collection manager at London Zoo believed that reports of a leopard in Kidderminster were well substantiated. He found them more convincing than reports of the beast of Bodmin which received considerably more press attention. Nick Morris photographed what appears to be a black leopard in Kidderminster in 1992. This and other sightings in the Cotswolds earned the elusive cat various nicknames, including the Beast of Gloucester, Beast of Inkberrow or Beast of any village or town where it had allegedly been seen

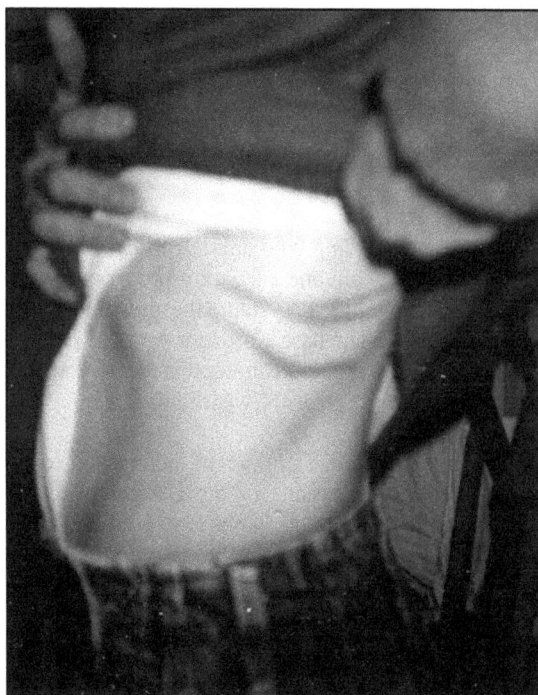

The Express and Star reported on 1 February 1994 that a black cat, the size of a Great Dane, had attacked a woman in a churchyard at Inkberrow. A man named Nick Dyke was trying to bait the animal with dead chickens. It responded by lashing out as his female companion before running away. On 21 April 1994 Sally Dyke appeared on a television programme called 3D. She said that when returning to check the bait they had been attacked and the cat ripped through her waxed jacket to scar her chest. The incident was not reported at the time, in December 1993, and Sally treated the wound herself and took

This picture, in which Sally Dyke shows off her wounds, was amongst an archive, formerly belonging to a magazine called *The Crypto Chronicle* which was purchased by the CFZ in 1995

antibiotics. Marks were visible on her chest at the time of the programme.

According to the BBC four teenagers camping near Bisley in Gloucestershire in 1994 were chased by what appeared to be a large cat.[13] It was larger than a Labrador.

Straying over the Worcestershire border into Herefordshire a cow belonging to farmer Norman Edwards of Pennbridge, near Leominster, was found with deep slashes on its rump according to a *Daily Mail* report on 2 May 1995. The local vet was certain that it had been attacked by a big cat. Two sightings of wild cats were recorded in the 1997 Mammal report of the Gloucestershire Naturalists Society.[14]

On 2 February 1998 Mr Lawrence Robertson, the Conservative MP for Tewkesbury who is still in post at the time of writing, interrupted a parliamentary debate about big cats in Norfolk to say that on several occasions farmers in his area had shown him photographic evidence of attacks on farm animals which he did not believe could have been carried out by anything other than big cats. They included lambs with their heads ripped off and fang marks of a width that no dog could produce.[15] Mr Robertson informed me in an email on 22 December 2009 that he remembered the debate but nothing else on the topic had come to him since then.

On 8 October 1999 the Forest of Dean and Wye Valley Review reported that in August 1998 an elderly lady from Pillowell, a village on the south-eastern edge of the Forest of Dean, looked out of her window in Link Road late at night and saw a huge panther like creature sitting outside her front gate. On 22 May 1999 the Birmingham Post reported several sightings of a large, black creature with a long cat-like tail across mid-Worcestershire. The most recent tracks, found at Ombersley, had been identified by Bob Lawrence, head warden of the West Midlands Safari Park as those of a dog or large fox. Despite this Mr Lawrence expressed interest in seeing any further evidence relating to big cats.

A lady saw a large black cat cross the road near Five Acres garage in Coleford, which is the administrative centre of the Forest of Dean, at around half past eight on a morning in September 1999. On 10 September 1999 the Forest of Dean and Wye Valley Review said that a serving policeman had contacted them to say that he had reported sighting a puma in 1996 whilst on duty. Both he and his partner saw the animal clearly.

There are now various big cat groups in Britain who receive details of sightings from members of the public via email, correspondence or telephone. In 2002 one of these groups, the British Big Cats Society recorded one thousand and seventy seven sightings nationwide, including fifty three in Gloucestershire. In the period between January 2003 and March 2004 this had risen to two thousand and fifty two nationwide, with sixty four in Gloucestershire and nineteen in Worcestershire.[16] Between April 2004 and July 2005 the national figure was two thousand one hundred and twenty three with only Devon and Yorkshire reporting more sightings than Gloucestershire.[17] The Big Cats in Britain Group were notified of six hundred and seventy five sightings in 2007, thirty one in Gloucestershire and three in Worcestershire. Since 2006 the Big Cats in Britain Group has published a yearbook, listing all sightings reported to them. They also publish a magazine, with various reports and articles. The number of sightings alone suggests that the subject should be taken seriously.

If big cats are living in the wild in Britain one has to ask where they originated. Any animals released in 1976 following the passing of the Dangerous Wild Animals Act must have bred to account for more recent sightings. Due to the current regulations, and increased communications, it is unlikely that large animals could now escape from zoos and safari parks without the loss being reported. In 1969 when it was suggested that licensing of wild animals was required Mr Merlyn Rees, the MP for Leeds South, boldly said that there was no reason to think that escapes of dangerous wild animals were numerous enough to necessitate legislation.[18]

This statement was proven false in 2001 when DEFRA released a list of big cats known to have escaped in Britain.[19] There were twenty eight cases, reported between 1975 and 2001. All but four of the animals were recaptured, shot or found dead. The exceptions were a Caracal, which escaped from Kent in 1980, a Lynx which escaped from Norfolk in 1991, an Asiatic Golden Cat which escaped from Somerset in 1997 and a jungle cat reported from Shropshire in 1993. The full list is summarized below.

- A leopard escaped on 1 January 1975 and was recaptured in Kent four days later.
- A clouded leopard escaped on 1 August 1975 and was shot in Kent on 1 January 1976.
- A puma escaped on 1 January 1980 and was recaptured in Inverness on the same date but not reported until 30 October 1980.
- A caracal escaped on 1 January 1980 and was shot in Kent on 22 November 1983.
- An ocelot escaped in Lancashire on 1 November 1981 and was shot on the same day.
- A jaguar escaped in North Wales on 1 September 1982 and was shot on 20 September 1982.
- A lion escaped in Norfolk on 5 January 1984 and was shot on the same day.
- A tiger escaped in Kent on 1 October 1984 and was shot on the same day.
- A leopard cat escaped on 1 January 1987 and was shot on 6 March 1988 in the Scottish borders.
  A Bengal leopard Cat escaped on 1 October 1987 and was shot in the Scottish Borders on 6 March 1988. This may be connected to, or a duplication of, the previous entry.
- A clouded leopard escaped in Kent on 22 October 1987 and was recaptured a week later.
- A puma escaped in Leicestershire on 1 October 1988 and died in an accident on the same day.
- A Bengal leopard cat escaped on 1 March 1988 and was shot in Devon on the same day.
- A leopard escaped in Kent on 1 June 1988 and was shot on 21 June 1988.
- A jungle cat escaped in Hampshire on 29 July 1988 and died in a road accident on 28 May 1989.
- A jungle cat escaped in Shropshire on 3 February 1989 and was found dead the same day from unknown causes.
- A lynx escaped in Norfolk on 1 January 1991. Its fate was unknown.
- A lion escaped in Humberside on 1 January 1991 and was recaptured the same day.

- A snow leopard escaped in Hertfordshire on 29 November 1994 and was recaptured the same day.
- A lynx escaped in Oxfordshire on 5 November 1996 and was recaptured on 28 November 1996.
- An Asiatic golden cat escaped on 1 September 1997 in Somerset. Its fate was unknown.
- A Bengal leopard cat was shot on 1 January 1987 on the Isle of Wight.
- A leopard cat was shot on 1 January 1987 on the Isle of Wight. This may be a duplication of the above.
- A Eurasian lynx was caught in London on 8 May 2001 and reported the following day.
- A leopard cat was shot in Devon on 22 April 1988.
- A leopard cat was found dead in the Scottish borders on 1 January 1990.
- A jungle cat was reported missing in Shropshire on 22 November 1993.

During the parliamentary debate on big cats in Norfolk in 1998 Elliott Morley who was then the Parliamentary Secretary for Fisheries and the Countryside said that the Ministry of Agriculture, Fisheries and Food was aware that sixteen big cats had escaped into the wild since 1977. They included lions, tigers, leopards, jaguars and pumas but only two animals were at large for more than a day. A leopard avoided capture for a week and the origin of a puma, found near Inverness in 1980 was unclear but it appeared to be tame. This information does not tally with that provided by DEFRA three years later. Only two of their twenty seven reported cases occurred after the debate. Additionally, as is evident from the list, most were not captured on the same day. Such discrepancies do not inspire confidence in the government's ability to record and disseminate accurate information.

Some researchers have suggested that the animals now seen could be descendants of beasts that escaped from Victorian menageries. We know that there was some trade in exotic animals during the nineteenth century. On 5 May 1855 the Bristol Mercury reported on the sale of beasts from the famous Wombwell's menagerie, five years after the death of George Wombwell, as follows:

> "Wombwell's Menagerie: On Friday the whole of the collection, vans, wild beast cages and other effects, part of the once celebrated Wombwell's Menagerie and lately known as George Wombwell's Show was sold by auction and fetched a mere trifle. The sale took place at the Novia Scotia Gardens, where last exhibited. The vans, five in number, nearly new and in good order and the stage van fetched from £7 to £10 each not the original cost of the springs. The beasts were, a jaguar, a leopard and leopardess, a nylgau, a hyena, a jackal, six monkeys, an alpine wolf, a raccoon, baboons, civet cat, some birds, a Russian bear, two beautiful Esquimaux dogs and, with the tilts, paintings, harnesses and usual paraphernalia, did not realise £200. The beasts, which were in very find condition scarcely found bidders. Bruin fetched £3 15 shillings, the monkeys, which were very choice sorts, from 5 shillings to 10 shillings and the two Esquimaux

dogs, 6 shillings and 4 shillings each."

If a suitable price could not be obtained at auction it is possible that owners of the travelling fairs would release their unwanted animals. For the puppies and kitchens of today substitute baboons and bears.

The argument against the existence of big cats in Britain is usually based on the likelihood of them being able to survive in the wild without being officially noticed. This is especially true in respect of towns such as Cheltenham, Gloucester and Worcester where there are many people and insufficient food for the cats although foxes have become adept at scavenging in urban areas and smaller creatures, such as rats, often go unnoticed. In rural areas there are often sufficient quantities of rabbits, hares, rodents and deer to feed a small population of wild cats who might live in areas rarely accessed by humans at night.

The Forest of Dean was the second largest royal forest in England, after the New Forest. It was well stocked with deer and boar for the privileged visitors. Now it is exactly the sort of location where big cats could conceivably survive and hide. Wild boar were reported as being present in the Forest of Dean for years before their presence was officially accepted. Sightings of big cats have been reported on a regular basis in and around the Forest of Dean since 1994. On 20 June 1999 the *Daily Mirror* reported that a sheep was found with its head ripped off and quoted Ranger Eric Pritchard as saying that a dog couldn't have carried out this sort of attack. Scratch marks that could have been made by a puma were found on a tree near the sight of the incident which was close to the Mallards Pike beauty spot in the heart of the forest.

Reports of big cats in the Forest of Dean were given credence in January 2009 when the Forestry Commission responded to a Freedom of Information request by revealing that two big cats were caught on camera by rangers who were filming as part of a three year long deer survey. Using thermal imaging cameras they had spotted the animals on two separate occasions in different parts of the forest. Commission spokesman Stuart Burgess said the sights had been confirmed by experienced rangers who were very unlikely to mistake deer for big cats.

The first sighting was at the outskirts of Churchill enclosure, east of Parkend in February 2002. The second was on the southern slopes of Staple Edge in March 2005. Deputy surveyor Rob Guest, who was present at both sightings, was quoted in several national newspapers as saying that, in his eighteen year career with the Forestry Commission, he had not seen cats like this before. The images were not recorded.

The Sun in its usual blunt style, under a heading "The Tooth is Out" on 7 January 2009 accused government officials of hushing up the truth. It drew a less than helpful comparison between puma, panther, lynx and moggie. Whilst this type of publicity can hinder serious researchers the newspaper is quite right to ask questions about the officials. It quoted DEFRA as saying that these were isolated incidents with no evidence of breeding groups. Neither DEFRA nor the Forestry Commission explained why the information was not made public at the time. Were the rangers perhaps sworn to secrecy by their superiors? There have been suggestions that farmers might attempt to claim compensation for livestock lost to big cats if the existence of the latter were to be accepted. In addition any Government acknowledging the presence of big cats in Britain would have to address perceptions of the risk to public safety and have a policy on hunting.

In general local media are more likely to take a serious approach to the issue of big cats in Britain than their national counterparts. Rick Minter is an environmental consultant who searches for big cats in Gloucestershire. He has been interviewed several times by BBC Radio Gloucestershire and answered questions from listeners. He also gives talks and presentations in various locations, often in association with Frank Tunbridge who has been investigating big cats in Gloucestershire for around twenty-seven years. Mr Tunbridge has had several sightings of his own. In an article in the *2008 Big Cat Yearbook* he said that fifteen people out of every sixty questioned in rural areas had either seen a big cat or knew someone who had. On 20 August 2009 the BBC Gloucestershire website published some footage that Mr Tunbridge had recorded during his night vigils in search of the cat. These included deer and dogs but, as yet, no cat. Three years earlier the same website had hosted a video in which Mr Tunbridge spoke about his experiences with big cats and explained his theories about their origins.

In 2006 the Cotswold Water Park played host to a conference on big cats. Frank Tunbridge was one of the speakers, saying that he often received several phone calls every week from witnesses. A summary of the event, by Rick Minter, is available on the Big Cats in Britain website. Of interest to the counties covered in this book is the suggestion that at least two estate owners in Gloucestershire wanted the presence of big cats on their land kept secret and the revelation that Gloucestershire Police made an off-the- record admission that they had removed the body of a big cat killed on a road in Gloucester in 2000. Now there may be good reason to keep the presence of the cats secret, especially in urban areas where panic could easily spread, but concealment goes against the transparency committed to by public servants whose wages, lest we forget as many officials do, are funded by the taxpayer. The Ministry of Defence has also been accused of concealing the presence of big cats in Britain. On 31 January 2008 the Northern Echo reported that two big cats were said to have been killed on a road by motorists, one of whom worked for the Ministry of Defence at RAF Flyingdales on the North Yorkshire Moors. Mark Fraser, of the Big Cats in Britain Group, said that a witness came forward to say that the body of one of the animals was seen on the base. The Ministry of Defence denied this. Given the unexplained seven year silence of the Forestry Commission there is reason to suspect that some officials may know more than they wish to reveal.

Between 2001 and 2003 the Rural Development Service investigated twenty two cases of alleged livestock predation by big cats reported to them where there was potentially hard evidence. These were in the following counties: Buckinghamshire, Cornwall, Devon (three times), Essex (three times), Kent, Norfolk, North Somerset (twice), North Yorkshire, Oxfordshire, Shropshire, Staffordshire, Suffolk, Surrey (twice), Warwickshire, West Midlands and Wiltshire. In two cases a field visit was conducted. This involved a trained wildlife biologist examining the evidence.[20] The presence of big cats was not confirmed in any of these cases. In response to a later Freedom of Information request DEFRA published a list of dead animals where they had carried out post mortems in their laboratories in order to identify signs of predation by big cats.[21] The vast majority were examined at the Aberystwyth laboratory. It is unknown where the cattle and sheep were killed but the analysis of the bodies shows that DEFRA were taking big cat sightings seriously in the early years of the twenty first century.

In Gloucestershire at least two big cat incidents were investigated by DEFRA. The first related to sightings of a big cat in Wickwar between December 2005 and July 2006 where the evidence was considered to be inconclusive.[22] The same conclusion was drawn in regard to a May 2007 sighting in Stroud.

The wildlife liaison officer for Gloucestershire Police, Mark Robson, is a believer in big cats. He has stated in several newspaper interviews, including those following the Freedom of Information release from the Forestry Commission, that Gloucestershire Police take sightings seriously and receives around a hundred reports from the public each year. Mr Robson, a civilian, is based in a suburb of Gloucester called Quedgeley which has had its own big cat sightings. It is a shame that the reported eagerness of this officer does not extend to other areas of the Police's remit.

In March 2009 I submitted a request for information on sightings to Gloucestershire Police, under the Freedom of Information Act. It was not responded to within the statutory time-limits. A response was eventually received, three days after the Information Commissioner wrote, on my behalf, to the force.

Amongst the myriad pointless government organizations that waste taxpayers money on a daily basis this Commissioner stands, almost alone, as someone prepared to champion the rights of taxpayers to see the documents that they are actually paying for. The veil of secrecy surrounding government has at last been lifted, albeit not completely. Whilst it would be naïve to think that the Police and other organisations will hand over all documents they are now compelled to release some information or face an independent investigation into their failure to do so. The power of the Commissioner is continually expanding although it is unlikely that he will ever be allowed to exercise his ultimate power to deprive a government organization from having the right to store personal data.

The Freedom of Information Unit within Gloucestershire Police, based at Quedgeley, informed me that they had received twenty seven big cat sightings in the period between 1 June 2007 and 6 March 2009, recording the date, location and description along with an incident number.

They stated that they only retain details for two years, most organizations go back further than this, but kindly enclosed a previous report covering the period 2004-2006. This had details of twenty six sightings. This makes a total of fifty three sightings in a forty five month period, nowhere near the annual century claimed by Mr Robson.

The full text of the letter from Gloucestershire Police in response to the Freedom of Information request is reprinted below.

"Gloucestershire Constabulary Freedom of Information request 2009.1848

On 6th March 2009 you sent an e-mail constituting a request under the Freedom of Information Act. Please accept my apologies for the delay in responding to your request; this is due to an exceptional number of requests being received recently. We are currently addressing this issue and hope to engage more staff to deal with Freedom of Information requests imminently.

1. I would be grateful if you could provide details of all reported sightings of incongruous animals and birds in Gloucestershire since records began.

Under the Freedom of Information Act 2000 s1, I can confirm that the Gloucestershire Constabulary holds some relevant information. The only information recorded by Gloucestershire Constabulary that would fit into the category of incongruous animals and birds, is the recording of possible "big cat" sightings. Detailed below are the details of all reported sightings of suspected big cats in Gloucestershire since 1st June 2007. The incident recording systems used by Gloucestershire Constabulary only retains incidents for 2 years, after which they are no longer accessible or searchable, I have however located a previous Freedom of Information request which details sightings of suspected big cats over a more historic period, which may of use to you."

Big Cat Sightings Since 1st June 2007 - to date

Incident No.	Date	Location	Description
98	16.11.07	Charlton Abbotts	Large Cat Sighted
187	09.10.07	Bisley	Black Panther
367	04.09.07	Coleford	Big Cat Sighted
65	09.07.07	A417	Large Black Animal
61	15.07.07	Aldsworth	Large Cat/Leopard Sighted
636	13.06.07	Great Walford	Large Cat Sighted
74	14.06.07	Ruspidge	Big Cat Sighted
560	20.05.08	Cinderford	2 Large Black Cats
526	25.04.08	Cleeve Hill	Large Black Cat
522	09.04.08	Westbury on Severn	Big Cat Sighting

378	14.03.08	A361 Lechlade	Black Cat Sighted
9	29.02.08	B4227 Ruspidge	Large Black Cat
175	16.02.08	Cam	Big Cat in Field
578	07.11.08	Cheltenham	Black Cat Sighted
198	16.08.08	Alderton	Multiple Big Cat Sightings
75	27.06.08	Aldsworth	Big Cat Sighted
387	07.01.09	Nailsworth	Large Black Cat
337	08.01.09	Bromsberrow	Huge Black Cat
238	21.01.09	Deerhurst	Black Cat Sighting
86	13.03.09	Lower Wick	Very Large Black Cat
218	01.04.09	Cheltenham	Big Cat
168	26.04.09	Bishops Cleeve	2 Large Black Cats Seen
214	07.05.09	North Nibley	Wild Large Cat
204	08.06.09	Aston on Carrant	Very Large Cat
291	10.06.09	Longborough	Large Cat
498	29.06.09	Aschurch A46	Lion Sighting

Previous FOI request Big Cat Sightings 2004 - 2006

Incident No	Date	Location	Description
55	19.01.04	Cheltenham	Big Cat
131	24.03.04	Forest of Dean	Black Panther
207	29.04.04	Forest of Dean	Big Cat Seen
334	03.05.04	Sherston	Big Black Cat
134	10.05.04	Toddington	Black Cat Sighting
592	20.05.04	Gloucester	Black Cat Sighting
376	01.06.04	Forest of Dean	Big Cat Sighting
524	27.01.05	Slad	Black Panther
579	05.02.05	Frampton on Severn	Panther Sighting
199	20.03.05	Prestbury	Big Cat
218	22.04.05	Horsley, Stroud	Puma Sighted
276	01.06.05	Cirencester	Black Cat Sighting
433	27.07.05	Stow on the Wold	Large Black Cat
605	01.08.05	Barnett Way, Glos	Big Black Cat
587	02.08.05	A419 Cirencester	Black Panther
531	02.09.05	Cirencester	Black Panther
154	28.09.05	Chedworth	Large Cat
160	28.09.05	Chedworth	Panther Sighting
244	03.10.05	Cirencester	Black Cat
338	05.11.05	Cirencester	Big Black Cat
293	16.11.05	Yanworth	Black Cat

| 598 | 18.11.05 | Lydney | Panther Sighting |
| 230 | 22.11.05 | Ledbury | Wild Cat |

I hope this information is useful.

If you are not satisfied with this response or any actions taken in dealing with your request, you have the right to ask that we review your case under our internal procedure. If you decide to request that such a review is undertaken and following this process you are still unsatisfied, you then have the right to direct your complaint to the Information Commissioner for consideration.

If we can be of any further assistance please do not hesitate to contact me."

The most interesting of these reports is the one about a lion in Ashchurch which is a couple of miles outside of Tewkesbury. When someone reports a large black cat there are other animals, such as dogs and domestic cats that could conceivably explain the sighting. It is more difficult to find a specimen of British wildlife that could be mistaken for a lion. Hoaxing is also unlikely when a sighting is only reported to the Police and not the newspapers. I found no record in the local or national media of the incident. However there was an article on the BBC's website on 21 May 2009 which said that two NHS workers at a call-centre in Bradley Stoke had reported a lion to Avon and Somerset Police two days earlier. Forty one days and forty one miles separate the two incidents.

West Mercia Police, who cover Worcestershire, responded more promptly to the same Freedom of Information request. They stated that it would be too costly to extract all the information from their database which kept data for more than two years. It is worth reprinting their reply, if only as an exercise in proofreading and the inability of the government to communicate clearly and accurately.

"A search of the West Mercia Operational Information System by our Operations Department using the suggested search codes of Animal problems and suspicious circumstances produced the following results.

	North Worcestershire	South Worcestershire
Animal Problems	997	1010
Susp Circs	24566	25715
Total	25563	26725

To read all these incidents and extract the relevant data from the identified big cat sightings would grossly exceed the fees limit. Therefore, the cost of providing you with the information is above the amount to which we are legally required to respond i.e. the cost of locating and retrieving the information exceeds the 'appropriate level' as stated in the Freedom of Information (Fees and Appropriate Limit) Regulations 2004.

In accordance with the Freedom of Information Act 2000, this acts as a Refusal Notice for your request

In accordance with Section 16 of the Act I have a duty to provide advice and assistance in relation to your request.

1.   I conducted a free text search for the period from 2002 to date using the following terms: Panther, Cat, Puma, Jaguar, Tigers, Lynx, and Lion. This search produced 4338 results, even allowing a minimum of 1 minute to examine and read each record for its relevance to Big Cat sightings would involve in excess of 72 hours additional work.
2.   Please find attached some details of sightings for North Worcestershire which was obtained in February 2007. I hope you find this data useful for your project, but please be aware that it is not comprehensive and should be viewed as minimum figure.

Every effort has been made to ensure that the information provided is as accurate and comprehensive as possible

Your attention is drawn to the below which details your right of complaint.

Should you have any further enquiries concerning this matter, please write or telephone the Information Compliance Unit quoting the reference number above."

I decided that it was not worth pursuing the tantalising 4438 possible sightings as the Freedom of Information Act does accept cost as a valid reason for not providing information. Whilst organisations should consider allowing individuals to take their own copies of information held it is unlikely that the police would allow unsupervised access to the database.

I could have asked why the police are increasing their number of civilian employees if they are not prepared to assist the public. I might also have taken issue with their searching techniques since it was not my suggestion to use the criteria of animals and suspicious circumstances.

One wonders how else panthers and lions would appear on a police database apart from reported sightings. People complaining about television documentaries perhaps or referring to public houses that are named after lions. Or maybe there is a gang leader known as the panther in the underworld of Worcestershire.

It appears to be a pointless exercise for the police to spend time cataloguing big cat sightings if they are never going to release it to the public. Is there some internal purpose, hinted at by the alleged removal of the carcase in Gloucester that explains their collection of the data? The previous information from North Worcestershire obtained in response to a request from someone else in January 2007, which presumably was cost-effective to supply, listed twenty seven sightings since 2000 and stated if officers had been deployed to investigate or not. Just two resulted in deployment. One was on 19 November 2000 at 15:09 in Moor Hall Lane, Stourport.

The Police searched the area and provided information to the nearby West Midlands Safari Park about the sighting of a black panther. They have said that they did not respond to such reports unless it was a current live sighting or there was a threat to life. In the context of the times when preventing and detected crime comes second to cost effectiveness and victims of reported crime do not always get a visit from the constabulary, it is perhaps understandable that the Police do not consider it productive to follow up sightings reported several days after the event. However this does contradict the stated policy of other forces. The second investigated sighting was in Kidderminster.

The full list is as follows:

18 January 2007	Past Shermock Court	A large cat the size of a springer spaniel was seen	No deployment
5 September 2006, 06:00	Chaddesley Corbett Kidderminster	Saw big cat in a field four days ago.	No deployment.
25 July 2006, 22:00	Churchill Kidderminster	Last night saw big cat middle of the road	No deployment
28 July 2005 19:00	South of Bromsgrove	On a train saw puma	No deployment
8 November 2004, 13:00	Stoke Prior, Bromsgrove	Saw black panther on a bridge	No deployment
4 October 2004, 18:00	Alvechurch	Sighting of black panther	No deployment
10 June 2003	Barnt Green	Saw puma on railway track	No deployment
20 August 2003 20:50	Hanbury	Saw big black cat in hedgerow	No deployment
30 August 2003, 15:14	Foxholes/Stourbridge Road	Saw big black panther	No deployment
21 April 2030 09:21	Stourport	Sighting of big black cat	No deployment

1 March 2003, 11:00	Beaulieu Close Kidderminster	Saw a big black cat	No deployment
2 February 2003 13:20	Broom Lane Clent	Saw black jaguar or panther	No deployment
12 February 2003	Habberley Valley, Kidderminster	Saw a cat (panther or cougar type)	No deployment
19 October 2002 10:00	Kinlet, Bewdley	Saw a very large black cat (panther)	No deployment
9 February 2001 07:59	M5 between Junctions 3 and 4	Saw black panther crossing a bridge	No deployment.
15 February 2001 13:20	Tennyson Way, Kidderminster	Saw big black cat up a tree.	No deployment
19 March 2001 10:30	M5 Gloucester junction	Saw large black cat in a field	No deployment
10 May 2001 01:12	Trimpley Drive, Kidderminster	Saw big black cat	No deployment
3 August 2001	Bigbury Lane, Stourport	Saw big black cat.	No deployment
9 August 2001 12:49	M5 J4A Overbridge	Saw black panther on the bridge	No deployment
16 November 2001 15:00	Marlbrook, Bromsgrove	May have seen a puma in a field.	No deployment
23 December 2000 11:20	Trimpley Lane Bewdley	Saw black panther jump into tree	No deployment
15 December 2000 07:55	Weaton Hill, Redditch	Saw rather large black cat.	No deployment
19 November 2000, 15:09	Moor Hall Lane, Stourport	Saw black panther	Area search, info to safari park
26 September 2000	Western Road Hagley	Saw puma in a garden	No deployment
15 September 2000	Farm Close, Kidderminster	Saw black panther twice	No deployment
4 September 2000	Cobham Road Kidderminster	Saw puma run into woods	No deployment
1 September 2000	Goldthorn Road Kidderminster	Saw large black cat in a garden	No deployment
30 August 2000 01:51	Spennells Valley Road, Kidderminster	Saw black panther in middle of road	Area search, no sign or evidence.

A couple of these sightings occurred on, or near, railway tracks. Foxes are known to use such tracks and it is reasonable to assume that big cats would do the same.

In 1992 I saw a fox, in broad daylight, saunter along the station platform at Witton in the West Midlands. There are, tragically, many deserted railway lines and those that remain in use rarely carry trains between midnight and five am.

Avon and Somerset Police which covers parts of Gloucestershire published, on their website, a response to a 2007 information request made under the Freedom of Information act.[23] The tone of this reply is much more friendly, and helpful than those of the neighbouring forces. The nineteenth, and last, of the sightings released was from Cromhall where three sheep were reported killed, possibly by a big cat, on 31 July 2006.

Not all sightings are officially reported to the police. Newspapers, most of which now carry online archives for the last three years, magazines and books contain details of numerous other sightings. Some witnesses contact the Centre for Fortean Zoology or other research groups.

There are also websites which contain reports and host forums which attract further reports from readers. *Fortean Times*, which continues to be the world's leading paranormal magazine, regularly publishes surveys of big cat sightings compiled from newspaper clippings submitted by readers. In considering the number of these reports made to various organisations, and the diversity of locations, it is sobering to realise that they constitute a small percentage of sightings. Those people who see what they believe to be a big cat have five options.

- They may decide to keep silent due to fear of ridicule, a desire to avoid publicity or because they do not realise the significance of the sighting.
- They may discuss the sighting with family and friends.
- They may report the sighting to the Police and/or other government agency.
- They may report the sighting to a research group, such as the Centre for Fortean Zoology, or a specialist magazine such as *Fortean Times*.
- They may tell the national or local media.

We have no way of knowing how many exercise the first and second options. Sometimes the silence is broken years after the event, perhaps following publicity of another sighting. Whilst valuable to researchers the trail is no longer there to follow.

Thanks to the Freedom of Information Act we can get some idea about reports made to the Police and these appear to be quite significant. Additionally the Police may themselves make candid comments to the media. Usually they reveal more to local newspapers although this is noticeably less common following the passing of the Freedom of Information Act.

The other significant agency to which big cats might be reported is the Department for the Environment and Rural Affairs which has had several expensive name changes in recent times, all funded by the taxpayer. DEFRA do not routinely record reports of big cat sightings.

The bulk of information about big cat sightings available to the public comes from the media although there is no guarantee that a reported sighting will be published. According to reports published in local and national newspapers the following big cats may be living wild in Gloucestershire and Worcestershire.

Leopard, *Panthera pardus*.

This species has short legs and a long body. They can move at up to thirty six miles per hour and although they were resident in Britain during the mid-Pleistocene, in historical times they have never been officially resident in the wild in Britain.

Three were given to Henry III sometime around 1235 as a gift from Emperor Frederick II, the Holy Roman Emperor, who had married Henry's sister. They were kept in the tower of London zoo. Black leopards are known as black panthers.

Leopard cat, *Prionailurus bengalensis*.
This is a small wildcat from South East Asia and India. They can be bred with domestic cats to produce a hybrid, called a Bengal cat.

Lion, *Panthera leo*.
After the tiger this is the second-largest surviving cat species. Coloration varies from light buff to yellow, red or dark brown. The hairy tail tuft is black and faint spots may be seen on the legs and underparts. Giant lions were present in Britain up to 13,000 years ago. They were also kept in the royal zoo from the thirteenth century onwards.

Lynx, *Lynx lynx.*
These animals vary in colour from medium brown to gold to beige-white and may have dark brown sports. They all have white fur on their chests, bellies and the insides of their legs. They have short tails and characteristic tufts of black hair on the tip of their ears. They were wiped out in Britain around 500 AD or about 1500 years ago. [**]

Puma, *Puma concolor*.
Also known as a cougar, mountain lion or panther. They have round heads and erect ears. The colour varies, with the coat usually being tawny. This is a purely New World species, although a relative, now long extinct, lived in Europe (including Britain) in prehistoric times.

Scottish Wild Cat, *Felis silvestris grampia*.
This is the only wildcat that officially exists in Britain. They resemble a muscular domestic tabby with the gait of a larger cat and have a thick ringed tail.

Do any of these or other exotic feline species exist in Gloucestershire and Worcestershire? It might be possible to fill a whole book with attempts to answer that question. The purpose of this book is to chronicle the Cryptozoology of the country. In respect of big cats I have chosen to summarize some of the incidents that have been placed in the public domain in the twenty first century.

[**] The extinction was originally thought to be much earlier but carbon dating on some bones, originally found in the 1920s and 30s, prove that some were living in this period. That research is in Hetherington, D. A., Lord, T. C. and Jacobi, R. M, "New evidence for the occurrence of Eurasian lynx *(Lynx lynx)* in Medieval Britain, *Journal of Quaternary Science*, 2005, 21, 3-8.

PREVIOUS PAGE (TOP) Melanistic leopard (*Panthera pardus*) also known as black panther (BELOW) Leopard cat (*Prionailurus bengalensis*) THIS PAGE (TOP) Female lion (*Panthera leo*) (BELOW) Eurasian lynx (*Lynx lynx*)

(ABOVE) Puma (*Puma concolor*) (BELOW) The Scottish race of the Eurasian wildcat (*Felis sylvestris grampia*)

Chedworth, not far from the road where Ty Gurr saw a big cat. Photo taken by author on 21 May 2010.

THREE

Panthers and Pumas in the Papers

The following list includes reports in newspapers, books, on websites and in magazines. It is not intended to be exhaustive. In most cases other media outlets published similar stories which I have not mentioned or referenced.

5 February 2000. Source: BBC website.
A spokesperson from Gloucestershire Police was quoted as saying that a collie cross-breed dog had been attacked by a big cat in the village of Longhope. In the second incident, in Newent, a man awoke to find claw marks, five millimetres deep, in his front door. Derek Day, 63, said he had heard a commotion during the night as something tried to get through the cat flap.

7 February 2000. Source: *The Sun.*
More details were provided on the attack on a pet collie in Longhope. The dog, Basil, was found by his owner Lin Byard with a four inch gash in his neck. Local vet Nick Clayton said that it had the hallmarks of a cat bite but was much larger than a domestic cat. He referred to other reports of panthers and pumas in the area. Four mutilated sheep had recently been discovered and it was feared that the cat might harm children. This is a common fear expressed in the tabloids with no supporting evidence.

11 February 2000. Source: *The Forest of Dean and Wye Valley Review.*
The newspaper interviewed Danny Nineham who said that a couple of trained tracker dogs would help catch big cats. Mr Nineham, who lives in Gloucestershire, has spent several years trying to find the cats and experienced his own sightings.

14 February 2000. Source: *The Birmingham Evening Mail.*
Former Midland Lion tamer Leslie Maiden said that he released a panther and cougar into the Derbyshire countryside during the 1970s. Bob Lawrence, head warden at West Midlands Sa-

fari Park was convinced that there were some big cats out there. He felt that the best evidence came from 1992 when a panther-like creature was filmed in the Inkberrow area.

The safari park, just off the A456, opened in 1973. Amongst other animals it contains cheetahs, lions, white lions, Bengal tigers and leopards. West Mercia Police have been known to call on experts from the safari park to help them look for big cats seen in other counties. One such incident occurred on 25 November 2003 when a main road was closed in Shrewsbury.

The Evening Mail also spoke to Mr Lewis Foley, who once kept lions at his home in Cradley Heath. He said that a friend of his set a panther loose in Nottinghamshire in 1974. A RSPCA spokesman said that it only became illegal to release wild animals into the countryside in the early 1980s.

21 February 2000. Source: *The Gloucestershire Echo*.
Danny Nineham believed that a leopard was responsible for savaging a ewe at Severn Springs near Cheltenham.

2 June 2000. Source: *The Forest of Dean and Wye Valley Review*.
A lady had found large animal dung, possibly from a big cat, at the back of her garden in Littledean. Danny Nineham collected samples.

16 June 2000. Source: *The Forest of Dean and Wye Valley Review*.
Danny Nineham was convinced that a leopard was loose in the Newent, May Hill and Longhope area. A woman who lived in a remote part of Longhope said that her large Doberman had suffered terrible injuries about a year ago. Mr Nineham had examined the dog and felt that the scratches were made by a large cat. This would appear to be a different dog to the one that was attacked in Longhope in January 2000.

23 June 2000. Source: *The Forest of Dean and Wye Valley Review*.
The newspaper published a photograph of a sheep with marks that suggested it had been killed by a big cat.

4 August 2000. Source: *The Forest of Dean and Wye Valley Review*.
Brian Elliott from Lydney saw a big black cat on a rock near Cannop.

10 August 2000. Source: *The Birmingham Evening Mail*.
Mr Roger Kordas from Stocks Lane, Leigh Sinton in Malvern was walking his dogs in a wheat field when a large muscular animal walked past him at a distance of ten to fifteen yards. He reported this to West Mercia Police and was adamant that he had seen a jaguar or leopard.

A police spokesman said that Mr Kordas then spoke to a local farmer who had lost some lambs. In recent years there had been several sightings around the River Severn and several people had reported animals being injured in Great Witley. Bob Lawrence had been called in several times to check animal prints after the sightings

11 August 2000. Source: *The Forest of Dean and Wye Valley Review.*
A parish footpath warden had discovered large cats hunting wild pigs in woods by farmland near Tideham. Kevin Fieldhouse said that he had tracked the animals in snow and seen them walking about. One was black and the other was fawn. Two other sightings were noted, one near Great Doward and the other at Parkend.

26 August 2000. Source: *The Daily Mirror.*
Following extensive media coverage of an alleged attack by a big cat on an eleven year old schoolboy in South Wales the Mirror interviewed Danny Nineham who claimed to have seen big cats frequently near his home in Lydney.

1 September 2000. Source: *The Forest of Dean and Wye Valley Review.*
Another interview with Danny Nineham. The article also noted a recent sighting from Susan Evans of Whitecroft who saw a large cat whilst walking her dogs in woodland near her home.

There have been several big cat sightings in and around the Forest of Dean. Photo taken by author on 31 May 2010.

5 September 2000. Source: *The Birmingham Post.*
A police helicopter, assisted by Bob Lawrence, had joined the hunt for a puma after it was seen in the area of Spennells Valley Road and Comberton Road, Kidderminster at 10:30 am the previous day. This would appear to be the incident reported to West Mercia Police.

24 September 2000: *The Sunday Mercury.*
The newspaper interviewed Guy Smith who had been searching for big cats in Britain for twenty years and had published a book, *Hunting Big Cats in Britain.* Following reports of pumas near his Shropshire home Mr Smith had constructed a steel cage to catch one. The trap was not successful. The article listed a selection of sightings in the Midlands since 1975. In Gloucestershire and Worcestershire these were:

- Inkberrow, 1992, the filming of a black leopard mentioned above.
- Kidderminster 1992, an Asian jungle cat was spotted.
- Coln Valley, Gloucestershire, 1999, a large black cat was spotted by a farmer.
- Newent, February 2000, the incident with the dog referred to above.
- Malvern, August 2000, the beast of Malvern was spotted.

26 October 2000. Source: *The Worcester Evening News.*
A farmer from Honeybourne was convinced that a big cat had killed two of his sheep. He discovered the remains of an eight month year-old lamb on 18 October and a second corpse two days later. Both were killed in the middle of an open field before being skinned and picked clean of their flesh.

Quentin Rose, an animal consultant who carried out post-mortems on dead livestock said that the sheep probably died of natural causes and were then stripped of their flesh by scavenging foxes or badges. However he claimed that an elderly black leopard or panther was at large in Worcestershire countryside and that it had been mating with a male, producing at least three cubs. Mr Rose passed away in 2002 but before that had found time to search for big cats in Tewkesbury and other locations.

24 November 2000. Source: *The Birmingham Evening Mail.*
An eighteen month old pony, Jenna, was attacked by an animal believed to be a puma. She was discovered with horrific injuries in a field off the High Street in Belbroughton near Stourbridge. A big cat had been sighted in Bromsgrove earlier that week. Vet Mike Overton said it seemed that a big cat had leapt on Jenna and ate her flesh.

2001. Source: *Fortean Times.*
Sightings in the following locations were reported.

- 3 February, Tibberton, near Newent, seen at eighty yards.
- Mid February, Shurdington.
- 23 February, M48 near Mathern.
- 8 March, Grange Village, near Newnham-on-Severn: three witnesses.

- Early April, Brockweir. Many lambs mutilated.
- May, Shurdington area.
- 1 May, Leckhampton.
- 8 May, Bishop's Cleeve, seen at fifteen yards.
- 10 May, near Crickley Hill, brown, lynx-like.
- 10 May, Down Hatherley.
- Mid-August, Brockworth, light brown alien big cat.
- 16 August, Standish Woods, near Stroud, "deafening roar" heard by five children.
- 20 August, B4008 between Quedgeley and Standish.
- 25 Sept, Bull's Cross.
- October, Wychbold.
- October, Warndon, four sightings.
- Early November, near Cirencester, between Coates & Tarlton.
- Early November, near Birdlip, seen at fifty yards.
- December, Bringsty Common, a large pawprint found.

11 February 2001. Source: *The Sunday Times.*

Paul Sieveking, the editor of *Fortean Times*, wrote an article in which he said that had been several sightings of big cats in Gloucestershire on both sides of the Severn. At two in the morning just before New Year the Cadmore Cat attacked barman Ty Gurr, as he cycled between Chedworth and Barnsley near Cirencester. The pronunciation of the victim's name, tiger, suggests a hoax. He is described as a handyman from New Zealand working at the Hare and Hounds public house in Chedworth. An internet search did locate references to a man named Ty Gurr in New Zealand and some residents stated in 2010 that they remembered him.

24 February 2001. Source: *The Worcester Evening News.*

Farmer Mike Sabin and his daughter-in-law Sarah had seen a large cat whilst feeding cattle in fields near Stocks Lane Newland on 23 February. This was just six months after the sighting by Mr Kordas, who lived in Stocks Lane.

6 April 2001. Source: *The Forest of Dean and Wye Valley Review.*

The newspaper published a photograph of a dead ewe that Danny Nineham believed had been killed by a big cat near Chepstow racecourse.

13 April 2001. Source: *The Forest of Dean and Wye Valley Review.*

A farmer at Brockweir, Ron Peacey, found several ewes injured by a predator. They were bitten on the neck and lambs were being carried off. He said that a vet from the Ministry of Agriculture and Fisheries, MAFF, had visited and claimed that a fox was not responsible.

9 August 2001. Source: *The Worcester Evening News.*

A black leopard was seen on 5 August by two fishermen at Broughton Hackett. It might have been the cub of the big cat that had been roaming Worcestershire for years. The fishermen, Brian and Anthony Wiggin were convinced that it was not a domestic cat.

10 August 2001. Source: *The Kidderminster Shuttle.*

A black leopard had been seen by Tom Jenkins in Gilberts End, Upton-Upon-Severn. A spokesman for West Mercia Police said that it would be difficult to tell how the animal would react if it was cornered or under pressure.

27 August 2001. Source: *The Gloucestershire Echo.*

A woman was driving between Quedgeley and Standish on the B4008 at 20:15 on 20 August when she saw a lioness roaming in a field. The previous Thursday, 22 August, five children from Stroud ran home crying with fear after hearing a loud roar whilst riding their bikes in Standish Woods near Stroud. In Somerset three separate emergency calls were made during the previous weekend to report a lioness in a field by the A38 near Churchill. The Sun claimed on 24 August 2001 that fifteen people had reported seen this animal. Acting Inspector Andy Stone, from Stroud Police, said that callers had probably seen a puma rather than a lion.

24 September 2001. Source: *The Birmingham Evening Mail.*

A farmer at Wychbold had seen a large black animal eating food from a refuse sack the previous Friday morning, 21 September. The following day a man living at Lineholt, near Ormbersley, about six miles from Wychbold found a large hole with animal droppings inside. Previously people had seen a panther along the River Severn near Malvern and on the outskirts of Worcestershire. Bob Lawrence said that, from the description, it could have been a black leopard or jaguar. However he felt that the droppings were left by a badger as they dig latrines.

2002. Source: *Fortean Times.*

Sightings in the following locations were reported.

- February, Bream in the Forest of Dean.
- March, Bishop's Cleeve.
- 7 April, Cirencester.
- May, Leonard Stanley near Stroud.
- 12 August, Bishop's Norton near Tewkesbury.
- 4 September, Bishop's Cleeve.
- October, near Stroud, ginger-coloured.
- 17 October, Gorsley, almost hit by a car at dawn.
- 5 November, near Bromyard, seen at night by four footballers in a car.
- 18 June 2002. Source: The Forest of Dean and Wye Valley Review. Beryl Ryan was walking her Alsatian, Indie, near Whitemead park when they saw a big black cat.

28 November 2002. Source: *The Tewkesbury Ad Mag.*

The Barlett family from Sycamore Farm in Maugersbury had to have a ram put down after its hind legs were savaged. They also found a lamb's decapitated body. Gloucestershire Police warned Cotswold livestock farmers to keep an eye on their animals. Danny Nineham thought that a dog was responsible.

2003. Source: *Fortean Times.*

Sightings in the following locations were reported.

- Forest of Dean, two sightings one with a photograph and a pawprint.
- Cheltenham, with a pawprint.
- Chepstow, several sightings including one where the cat was eating a guinea pig.
- Near Monmouth, two sightings.
- Near Aston Down.
- Between Newnham and Blakeney where a cat was hit by a car and left hairs on the bumper.
- Near Forthampton.
- Near Tewkesbury.
- Near Moreton.
- Hardwicke.
- Near Newland.
- Cranham, a mutilation.
- Pedmore, three witnesses.
- Cradley.
- Ankerdine Hill, the animal was described as half fox cub and half wild-boar. There are reasons for believing that this was not a big cat and it is discussed in a later chapter of this book.
- Near Droitwich. This animal had a long snout and was bald apart from a big red mane. It was similar to the animal seen on Ankerdine Hill.
- Hadley.

15 January 2003. Source: *The Worcester Evening News.*

Fifty six year-old Christine Carley, of Westbury Avenue in Droitwich saw a big cat patrolling in snow outside her window at 06:45 on 9 January. She said the creature walked past a sheltered housing complex and across her road towards fields. She found fresh paw-prints, two inches across without claw marks.

19 September 2003. Source: *The Cotswold Journal.*

Phil Haling saw a big cat the size of a German shepherd prowling around next to the Trooper's Lodge petrol station on the A44 near Moreton. It was dark brown with pointed ears and a long tail. The sighting was at 18:45 on 17 February. Colin Northcott, deputy warden at West Midlands Safari park, was quoted as saying that he often received reports of big cats in Gloucestershire and Worcestershire.

2004. Source: *Fortean Times.*

Sightings were reported in the following locations.

- Tewkesbury.
- Abbeydale.

- A444 near the A424 junction.
- Fairview, a pawprint.
- Wychbold.
- Droitwich.
- Elmbridge.
- Between Feckenham and Hanbury.

4 January 2004. Source: *The Sunday Mercury*.
The National Farmers Union was encouraging members to report sightings of big cats. The locations of latest sightings in Worcestershire included Wychbold, Droitwich and Elmbridge.

29 January 2004. Source: *The Worcester Evening News*.
A cat had put a disabled woman in hospital. Linda Meredith from Stourport-on-Severn was in the bedroom of her Clarewen Avenue home when a cat pounced on her and took a large chunk of flesh from her leg. The report does not suggest that the cat was more than a domestic feline but such behaviour is not normal for tame felines.

19 August 2004. Source: *The Worcester Evening News*.
A large black cat had been seen by Anne Kirk and her sixteen-year-old daughter, Ettie. The animal was prowling on land near the Kirk's Ditchford Bank home between Feckenham and Droitwich. Ettie's pet pony had tried to escape her paddock twice in the previous two weeks, perhaps because she was disturbed by the cat's presence. The cat was described as having a big flat head, a long body, short legs and a long tail with round ears. Danny Nineham said it sounded like a black leopard. Residents in Hanbury and Wychbold had reported a number of big cat sightings in recent years.

25 August 2004. Source: *The Malvern Gazette*.
Jeffrey Brittain, from Selly Oak in Birmingham was fishing with his nephew in Stoke Prior when he saw a big black Cat. It was about 02:20. The cat was bigger than a Labrador with rounded ears.

2 September 2004. Source: Wotton-under-edge community website.

A poster on the website stated that several people had seen a large black feline in the vicinity of Hack Hill on the previous day.

2005. Source: *Fortean Times*.

Sightings were reported in the following locations.

- Forest of Dean.
- Gloucester.
- Woodmancote.
- Badgeworth and Leckhampton, several sightings.

- Five Valleys, several sightings.
- Cirencester, CCTV footage.
- Watermoor.
- Stratton.

11 May 2005. Source: *The Gloucestershire Echo.*

A big black cat had been seen wandering around Woodmancote and Leckhampton. Hucclecote pensioners, Mary and Jim Alison believed that it had devoured their pet cat, Tigger. Seventy year old John Blenkinsop said that he had seen a big cat on Cleeve Hill.

4 June 2005. Source: *The Stroud News and Journal.*

George Hearn was driving his taxi on the road between Aston Down and the Ragged Cot at half past midnight on the previous morning when a black cat glided across the road in front of him. It was about the size of a Labrador.

23 June 2005. Source: *The Wiltshire and Gloucestershire Standard.*

Watermoor resident Pauline Saunders had seen a feline creature the size of a dog prowling the street. Through her half-opened door in the early morning she watched it stroll along a row of cars under the light of street lamps before disappearing into the darkness. Danny Bamping, founder of the British Big Cat Society, was quoted as saying that there had been several sightings in Watermoor.

30 June 2005. Source: *The Wiltshire and Gloucestershire Standard.*

The Chief Executive of the Stratton District Council, Bob Austin, had seen a big cat in Stratton. This was near the children's' playing fields on the corner of Grange Court. Mr Austin described the animal, seen from a distance of approximately seventy five yards, as about the size of a Labrador.

20 July 2005. Source: *The Stroud News and Journal.*

The newspaper published a feature on big cats. This included a list of several sightings, including one from seventy three year-old retired policeman Roy Harvey from Amberley, who saw what looked like a black panther in his garden early one morning. It was bigger than a domestic cat with black to brown fur.

26 August 2005. Source: *The Dursley Gazette.*

A Wickwar famer lost one of his pedigree sheep to a big cat. John Terrett arrived at Osbourne's farm to find the dead animal lying on the ground. Mr Terrett said that he had seen a big cat about four years earlier and, again, at Christmas two years previously. The loss of the sheep was reported to DEFRA and to Farmers Weekly. DEFRA said that they took sightings seriously but there was no evidence that big cats were loose in the countryside. The sighting was also discussed on the local community's website.

2 September 2005. Source: *The Wiltshire and Gloucestershire Standard.*
The newspaper proclaimed on its front page that an alien big cat had been captured on CCTV. In the footage the black cat is seen walking towards a zebra crossing in the centre of Cirencester. It was taken by security cameras. Danny Bamping said he was one hundred percent sure of the image's credibility. He said that around sixty percent of all sightings reported were of black cats, thirty two percent were of brown or sandy coloured ones which they believed to be pumas, cougars or mountain lions and another six percent were lynx. The British Big Cats Society estimated that around a third of all sightings were not big cats. The CCTV footage and a pawprint in the area were reported to Fortean Times. The Cirencester News also reported on the CCTV footage and said that a pensioner had seen a panther-like creature in a field from a distance from about two hundred yards.

16 September 2005. Source: *The Wiltshire and Gloucestershire Standard.*
Big cat footprints were found in Coates, two miles from where the Cirencester CCTV footage was taken. However, it appeared that the footprints dated back to earlier in the year.

26 October 2005. Source: *The Wiltshire and Gloucestershire Standard.*
Mark Robson had been appointed as the Wildlife Liaison Officer for Gloucestershire Police. He claimed that the Police had received several reports recently from Cirencester and South Cerney. Forty sightings were reported in the Cotswolds in the previously twelve months but Mr Robson felt that many people logged onto the big cat website first. He said the Police were aware of two species, black leopards and pumas. The black leopards were more commonly reported. Although a civilian Mr Robson had similar powers to a uniformed officer. He said that he had been investigating big cat sightings for eight years and was looking for evidence, such as DNA, from the scene.

2006. Source: *Fortean Times.*
Sightings were reported in the following locations.

- Pauntley where a pawprint was found.
- Evesham.
- Near Broadway Tower where a big cat was seen attacking sheep.

15 February 2006. Source: *The Gloucestershire Citizen.*
A large black panther was seen lazing in the sun on 10 February. Motorist Dave Long, aged fifty nine, from Kings Stanley near Stonehouse had pulled into a lay-by on the B4215 Highnam to Newent Road when he saw the creature two hundred yards away from the vehicle. He called the Police at about 12:30. Officers went into the woods and Mr Long saw rabbits and pheasants come out as if they were scared by something. A police spokesman said that they sent four officers but nothing was discovered.

6 March 2006. Source: *The Gloucestershire Citizen.*
A report on paw-prints found by Katharine Midgeley from Hartbury. She was also involved in a BBC Radio Gloucester interview on the subject of big cats. Katharine had photographed

some footprints she found in a field in Pauntley whilst walking her dog the previous year. She said that she heard a growl and her dog was spooked by something. She returned to the spot the next day and took the photos on her mobile phone. Frank Tunbridge described the prints as some of the best ground evidence that he had seen. The article on the BBC Gloucestershire website, which also gives an opportunity to hear an interview with the witness, prompted comments from several people who claimed their own sightings along with others who were more sceptical.[24]

17 March 2006. Source: *The Daily Telegraph.*
An article by Adam Edwards who claimed that his neighbour Simon Scott-White had collected several sightings of the Cadmore Cat before having one of his own. Edwards spoke to Terry Hooper, who was head of the exotic animals register. Hooper said there was a puma living in the valley.

23 March 2006. Source: BBC website.
An article on the BBC's website previewed the first Big Cat Conference in Britain. The Beast of Gloucester, first reported in 1993, was fifth in a list of top ten cats, behind the Beast of Bodmin Moor, the Exmoor Beast, the Leicestershire Big Cat and the Telford Puma. The article also quoted Mark Robson as saying that Gloucestershire Police had received forty to fifty sightings that year, mostly describing a creature with a dirty black coat. This is more than the seventeen sightings in 2005 released in response to my Freedom of Information request.

25 March 2006. Source: *The Gloucestershire Citizen.*
The newspaper published a photo of a big black cat sent in by a reader who claimed to have recorded the image on his mobile phone near Littledean.

7 April 2006. Source: *The Times.*
The Times reported on big cats in Aberdeen and took a long geographical detour to quote Mark Robson as saying

> "I believe these big cats do exist because there is just too much evidence. Besides the witness reports, we are getting a lot of farm animals being unusually killed."

20 April 2006. Source: *Worcester Evening News.*
Two large black cats had been seen at a business park near Bewdley. Kevin Millman saw he saw the animals prowling in a yard at his business unit at Lye Head, Rock.

3 May 2006. Source: *The Gloucestershire Echo.*
A cream-coloured animal, smaller than a fox but larger than a domestic cat had been seen around Whaddon. A couple from Cam Road wondered if it was responsible for killing two birds found dead in their garden. Their son thought the sighting was of a lynx.

12 May 2006. Source: *The Gloucestershire Citizen.*
West Mercia Police were investigating the sighting of a lynx-like cat at Unipart's premises in

Shinehill lane, Evesham. A security guard claimed to have seen the animal, for the third time, at 22:00 on 9 May. It was mousey brown with pointed ears and a tufted tail. The BBC report on this said that Worcestershire Police advised residents in the area to be aware of the sighting. Their spokesman was quoted as saying:

> "We are aware that there have been many, many sightings of 'large cats' in the South Worcestershire area over a long period of time without any of them being properly authenticated and are aware too that no members of the public have been injured in any way."

He added,

> "Indeed, if this were to prove to be a lynx, its first and natural reaction would be to run away from people."

20 May 2006. Source: *The Birmingham Post.*
An article on big cats, concentrating on Warwickshire but referring to the Evesham sighting mentioned above and one by pensioner Terry Dallaway who said that he had spotted a panther-like cat by a canal towpath in Redditch.

22 May 2006. Source: *The Gloucestershire Echo.*
Terry and Celia Higginbotham had discovered huge paw-prints on their back patio in Moore Road, Bourton-on-the-water. Their neighbour, Shirley Warr claimed to have seen a big cat on a previous occasion and showed the plaster cast of its prints. The Higginbottams were unaware of her experiences before they found the paw-prints.

22 June 2006. Source: *The Gloucestershire Citizen.*
An animal looking like a black panther was spotted in Swan Lane, Stoke Orchard at 10:00 on Sunday 18 June.

29 June 2006. Source: *The Gloucestershire Citizen.*
Emrys Davies saw a black panther in a field when he was travelling to a concert in Trellech church with three fellow members of Chepstow Male Voice Choir. The field was near Itton on the B4293 road from Chepstow to Devauden. There was bright sunshine at the time and Mr Davies described the animal as black with small, pointed ears on a big head. The choir's pianist, one of the passengers in the car, confirmed the sighting.

17 July 2006. Source: *The Gloucestershire Echo.*
David and Alice Ross had seen a large black cat in a field behind their house in Ham Road, Ham near Cheltenham on 14 July. It was about the size of a roe-deer with a long tail and a pushed button-like face.

19 July 2006. Source: BBC website.
Frank Tunbridge wrote an article suggesting that a hide with a trip camera could record evi-

dence of the big cats. He referred to two recent sightings in Gloucestershire. The first was at Port Ham Nature Reserve. A medium sized dark grey cat, with black tabby or blotched markings was seen by a man walking his dog on 13 July at 21:45. The second was the sighting by Alice and David Ross noted above.

27 July 2006. Source: *The Birmingham Evening Mail.*

A visitor to Broadway Tower in Worcestershire called police to report a large black animal making strange noises. It was the second possible sighting that year and a lynx type creature had been seen at Evesham. Police said that none of the sightings were authenticated.

17 September 2006. Source: *The Wiltshire and Gloucestershire Standard.*

Science teacher Rob Carmichael was cycling down Sheep Street in Cirencester late at night when he saw a big cat that resembled a lion. It was about the size of an Alsatian and had a bushy tail. The British Big Cats Society said that they had received thirty sightings in the area in the last six months. Gloucestershire Police confirmed three big cat sightings in the North Cotswolds in the last two months, two near Broadway and one near Chipping Camden.

22 September 2006. Source: *The Wiltshire and Gloucestershire Standard.*

Frank Tunbridge was interviewed. He believed that some cats had escaped from Victorian menagerie and felt that the government should take the matter more seriously.

12 October 2006. Source: *The Gloucestershire Echo.*

A report on Margaret Doherty's unexpected encounter with a big cat. The sixty year old from Winchombe was walking a friend's dog across parkland near Sudeley Castle when the cat ran across the fields in front of her.

20 October 2006. Source: *The Gloucestershire Echo.*

The newspaper reported a sighting by forty year-old Frances Green, her sixth sighting. She came face to face with a black panther outside Moreton District Hospital. She thought it might be from the Sleeply Hollow Farm Park at Blockley. Farmer Tim Spittle had hand-reared a twelve month old leopardess, Raphia and her brother, on the park near Moreton-in-Marsh. He also kept a puma, jaguar and three tigers. When he applied for planning permission to build a new enclosure to house the big cats in 1998 Cotswold District Council refused the application and ordered the park to be closed. However Tim and, his wife were not prosecuted. The Independent On Sunday reported on 7 December 1997 that a lynx had previously escaped from the park and had to be shot. Villagers felt that there had been other escapes.

2 November 2006. Source: *The Forester.*

Two-eleven year old boys had fled from a big cat in Cinderford. Joe Tingle and Chay Maidment were playing football in a field at Cinderford Bridge at around 17:00 on 30 October 2006 when they heard noises in a tree nearby. Suddenly there was a crash and a black animal with huge green eyes stared at them from the undergrowth. The newspaper also said that Police were looking for information after a couple out walking reported another possible panther sighting in Bullo near Newnham at 16:00 on 29 October. The Cheltenham couple said they saw a big black animal, larger than a dog but smaller than a pony. A police spokeswoman, Zoe Young, was quoted as saying that sightings were dealt with by the wildlife crimes officer who mapped them. If this is true then surely that wildlife crime officer should be able to release that map, and sufficient details, to anyone making a relevant Freedom of Information request. So far they have not, to my knowledge, done so.

26 December 2006. Source: BBC website.
Brian Jones had discovered the carcass of a deer whilst walking his dog near Ruspidge in the Forest of Dean on Christmas Day. The bones had been stripped of flesh.

1 January 2007. Source: *The Gloucestershire Citizen.*
Big cats were driving urban foxes out of Cinderford according to Danny Nineham who said that he had seen a black panther in Cinderford High Street, a puma on the Hilldene housing and a leopard with her cubs in the town. He further claimed that thirteen dogs had been lost.

20 January 2007. Source: *The Gloucestershire Echo.*
A security guard told police that he saw a panther. Gloucestershire Police said that the incident occurred at 01:30 on 18 January at Wormington petrol station between Tewkesbury and Evesham.

8 February 2007. Source: *The Gloucestershire Echo.*
Fireman Peter Bishop came face-to-face with a black panther in Causeway Road, Cinderford. He was carrying a take-away Chinese meal home to his family when the cat walked out in front of him and casually strolled away. Two other sightings were referred to. On 23 January at 21:10 a driver from Lydney was heading along Valley Road towards the Bridge Inn when he saw a black panther-like animal cross the road and walk into Linear Road. On 1 February a big pale cat was seen near a hedgerow in Bream Road near St Briavels.

21 June 2007. Source: *The Gloucestershire Echo.*
Forty seven year old milkman Robert Brinton came face to face with a big cat in Railway Road, Ruspidge at 04:30 on 14 June. It was fighting with a domestic cat and ran off when he approached. A few hours later an off-duty policeman saw a large cat in the same spot. This is probably the sighting logged by Gloucestershire Police on that day.

28 June 2007. Source: *The Forester.*
Thirteen year old Jody Motterham told his parents he had seen a black panther in the woods. He said the sighting was in Railway Road, the same location as Robert Brinton, at 21:45 when he was riding back from army cadets.

30 June 2007. Source: *The Daily Mirror.*
Richard Hammond, of *Top Gear* fame, commented that the possibility of big cats roaming around Gloucestershire brought something special to people's lives.

5 July 2007. Source: *The Worcester Evening News.*
The newspaper asked if a lamb belonging to celebrity chef, Gordon Ramsey, had been killed by a big cat. The lamb was being grazed on the Hertfordshire estate of David and Victoria Beckham when it died. The reason for this story might lie in a reference to an article on big cats in *Smallholder* magazine. *Smallholder* was part of the Newsquest Media group who also owned the *Worcester Evening News.*

8 August 2007. Source: *The Cotswold Observer.*
A report on a sighting of a big black cat with two cubs near Sudeley. Following flooding in the Winchombe area it was suggested that the site on Sudeley Hill was the only area of raised land. The paper also said that experts believed that many of the big cat sightings were due to a growing population of Bengal cat colonies in Britain. These cats are hybrids of Asian leopard cats and domestic cat. They resembled a cross between a tiger and a lynx. What the theory doesn't explain is why so many sightings are of black cats as opposed to stripy ones.

4 October 2007. Source: *The Gloucestershire Citizen.*
Sixteen year-old Natasha Baker was walking her dog, 'Buster', with boyfriend Mitchell Townsend in fields behind their home in Western Way, Dymock. It was about six in the evening when they spotted a large creature following a hedgerow on the other side of the field. It was seven to eight feet in size, black, with a long tail.

10 December 2007. Source: *The Gloucestershire Citizen.*
Chris Harmer, chairman of Horsley Parish Council was featured in an article about proposals to install CCTV to film big cats. Earlier that year Mr Harmer had written about big cats in the winter edition of Nailsworth Council's *The Fountain magazine* which can be found online. One morning he was enjoying his usual morning walk when he saw a jet black animal about the size of a great Dane running into the woods. He phoned Stroud Valleys Project for advice and spoke to a lady who said that, a few days previously, she had seen an animal cross the road just before the Owlpen gate and the right turn to Horsley. She described it as a beige coloured cat. On her advice Mr Harmer reported the matter to the Police and was contacted by a wildlife officer who said that he took the sighting very seriously. Meanwhile Mr Harmer's wife saw a black cat in the field past Hartley Bridge. She thought it was a domestic cat until she saw its long thick tail with a rounded end.

Mr Harmer reported that the number of deer in the woods beyond Hartley Bridge towards Kingcote had become less numerous. He knew of two deer kills, one in his neighbour's garden and one at the Nailsworth end of Rockness. He came across a third one in Barcelona Lane which bore the hallmarks of a big cat kill; a bite to the neck and a hole where the killer had gone for the internal organs. He emailed pictures to Stroud Valley Projects and they were shared with big cat experts. *The Gloucestershire Citizen* also quoted a police spokesperson as saying:

"we believe that there is evidence of their existence in the county."

20 December 2007. Source: *The Gloucestershire Citizen.*

Frank Tunbridge commented that trip or automatic cameras could be used to record big cats.

31 December 2007. Source: http://blackpantheranimal.com
Andrew Lloyd reported that he had seen a black panther at 12.00 whilst driving on the Droitwich Road near Bradley Green. It was around two foot at shoulder length and had fangs with green eyes and a large tongue.

10 January 2008. Source: *The Gloucestershire Echo.*
Cyclist Graham Hill had discovered a sheep's carcass picked clean at Burnt Log, opposite the New Fancy Viewpoint in the Forest of Dean. Danny Nineham found the remains of other animals nearby but was unable to determine what had killed them.

31 January 2008. Source: BBC website.
Frank Tunbridge contributed an article which included a photograph of a paw print found by a lady in Stroud on 12 January. The marks were left approximately three quarters of a mile from the location of a deer found dead on 16 January. Mr Tunbridge felt that the hallmarks of a big cat kill were there. These included

- A lot of flesh consumed quickly.
- Most of the ribs bitten off close to the spine.
- Badly twisted broken neck and strips of skin torn from nose area.
- Drag marks.
- Skin and fur peeled back in a clinical way.

On 4 February Mr Tunbridge updated the article to say that a muntjac deer kill had been reported to him on 31 January. Photographs of both kills were posted on the website. The article prompted several comments from readers including Andrew Goodair from Naunton who had noted the following in his diary on 29 November 2007:

> "saw a wild cat today 95% certain. It clocked us and disappeared very quickly but for the next mile or so it was really eerie and felt like the eyes of the beast were following our moves. The Mrs felt the same."

This occurred on a circular walk from Bourton via the slaughters. The animal was about fifty metres from him and had a dark brown coat.

6 March 2008. Source: *The Gloucestershire Citizen.*
A man was walking his dog along Ruspidge Road in Cinderford at 00:30 on 29 February when he came face to face with a leopard. He said he had reported this to the Police and this does appear to tally with the information provided by them. Another man was reported to have seen the cat in fields at the back of his house by the old railway line. The proprietor of Lower Ruspidge Stores said that several of his customers had seen the big cat at night on Railway Road.

17 March 2008. Source: *The Gloucestershire Citizen.*

An interview with Forest of Dean district councillor, Alastair Fraser, who wanted an animal tranquiliser to be used in the region. Mr Fraser, a llama specialist, was lamenting the death of a wild boar short by a ranger but also said a tranquiliser could be used on big cats. The boar's demise led to the presence of his species in the Forest being officially recognized.

18 July 2008. Source: *The Gloucestershire Citizen.*
Sixty two year old Roy Eisley from Tredworth was driving along the A417 near the M50 junction at 22:15 on 13 July when he saw a two foot tall black cat with a curled tail running across the road in front of his vehicle. After narrowly avoiding the animal, which he described as a panther, Roy pulled into the Rose and Crown pub where a customer told him that there were two cats roaming the area. The report was copied to the This is Gloucestershire website and prompted the usual flurry of comments from readers. They included a retired gamekeeper, calling himself Adam, who posted on 18 July 2008. He claimed to have been following the panthers for eighteen years but refused to reveal their locations. There was also a claim that farmers were killing big cats and not reporting it.[25]

9 August 2008. Source: *The Gloucestershire Citizen.*
A big cat was seen chasing a deer in Linear Park, Cinderford, near Valley Road. It had a long tail, a small head and squashed nose and was jet black.

21 August 2008. Source: *The Gloucestershire Citizen.*
Frank Tunbridge wrote an article in which he said that the cats were living comfortably in the countryside and that there was sufficient deer for them to feed on.

14 October 2008. Source: *The Stroud News and Journal.*
Stroud resident Jenny Bailey had seen a big cat at the Eastcombe crossroads at 20:05 on 9 October. It was black and as tall as a table. This comparison is not very helpful as some tables are bigger than others.

22 October 2008. Source: *The Gloucestershire Citizen.*
Amberley residents Steven and Jane Mansfield discovered a deer ripped in half on Minchinhampton Common.

6 January 2009. Source: The Gloucestershire Citizen.
A report on the Freedom of Information revelation made by the Forestry Commission. It said that witness Rob Guest confirmed there were no signs of big cats in the Forest during a deer census in March 2008.

11 February 2009. Source: BBC website.
Andy Janik posted on a forum to say that he had seen a large black cat like animal that day stalking four deer in Avening Woods just outside Nailsworth. It was about the size of a Rottweiler with a very long tail.

20 February 2009. Source: *The Gloucestershire Echo.*
Two tree surgeons from Tewkesbury, John Vine and Nick Cole, saw a big cat on a dirt track near the Highgrove Estate in Churchdown, Gloucester at 11.30. They were working on a willow tree when they saw the cat emerge from a nearby thicket. It was described as the size of a Labrador with a glossy shimmering coat, a long tail and a head similar to a domestic cat but bigger. The Sun spiced this up by informing readers that the location was near Prince Charles's estate. Frank Tunbridge visited the scene, saying that he believed the tree surgeons had disturbed the cat with their equipment. Following the local media coverage

Bernadette Smith posted on the *This is Gloucester* Website on 23 February 2009 to say that large footprints were left in the snow at the back of the *Hatherley Manor Hotel* and Debbie, from Churchdown, commented that she had seen a lot of very large footprints in the snow on her patio the previous week.

23 February 2009. Source: *The Dursley Gazette.*
In the early hours of the morning a large black cat was seen on the A38, coming out of Stone cricket club grounds and crossing the road in front of a vehicle. It was between two and three feet long, about the size of a border collie, with a long stretched body and feline head. Frank Tunbridge believed that it was the same beast reported elsewhere in the area. A spokesman for the cricket club said that a similar animal had been seen the previous year.

25 February 2009. Source: *The Gloucestershire Echo.*
Frank Tunbridge was sitting in a lay-by on the A419 between Stroud and Chalford with his digital camcorder. He hoped to obtain footage of the cat which had been seen in the area.

28 February 2009. Source: *The Western Daily Press.*
An interview with Danny Nineham who said that the cats were becoming more urbanized. The article also claimed that Gloucestershire Police received more than one hundred reports of sightings each year.

27 March 2009. Source: *The Worcester Evening News.*
An article about big cats, referring to sightings near Inkberrow at the turn of the century.

30 March 2009. Source: *The Birmingham Evening Mail.*
A big cat was spotted near to Bromyard road in Worcester by two men who were walking their whippets and terriers. The men were named as Wilson Hunt and Issac Biddle. The British Big Cats Society sent representatives to the area but they were unable to confirm the sightings.

4 April 2009. Source: *The Worcester Evening News.*
Farmer Anthony Thomas, from Martin Hussingtree had seen a big cat near Claines. He said that his farm had lost several sheep and the previous October found a ewe whose head had been taken off. The newspaper quoted Karen Allison, a representative of the Big Cats in Britain group, as saying that the best place to see big cats in Worcestershire was on the Malvern Hills, especially the area around Little Welland.

17 April 2009. Source: *The Worcester Evening News.*
The big cat had boosted tourism as visitors were coming to the county to look for the beast. This is perhaps surprising as Worcestershire is not one of the hotspots for big cat sightings with far more reports coming from Gloucestershire.

1 August 2009. Source: *The Stroud News and Journal.*
Frank Tunbridge wrote an article in which he listed some of the sightings reported to him in 2009. These were:

- **4 April, 23:40.** A man in Down View, Chalford was locking up his house when he saw a large black cat in the road. It was as big as a Labrador but with longer legs. The following night the man's brother was scared by noises in the night.
- **30 May, 22:00.** A woman was walking a beagle puppy in Penn Woods on the Cotswold Way when the dog stopped. She looked back to see a Labrador-sized black cat. It had orange eyes.
- **1 June, 04:45.** A man living in Butterrow Lane, Stroud got out of bed to use the toilet and looked into a neighbouring garden where a large black catlike animal was standing.
- **2 June.** A boy named Sam at Cam Hapton School in Cam looked out of the classroom window to see a large black cat. It ran around a corner, knocking over a flowerpot.
- **6 June, 08:15.** A member of the staff at the Aston Downs Car boot site, just off the A49 was starting to set up the site when he saw a large Labrador sized but skinny big cat come out of cover and trot over the hedge where it watched him for a few minutes before disappearing into the undergrowth.
- **18 June,** on the A417 between Ampney Street and the Red Lion at Poulton. A man and his wife were driving from Cirencester to Lechlade when they saw a large black cat which had either a white chicken or domestic cat in his mouth.

28 August 2009. Source: *The Gloucestershire Citizen.*
The newspaper claimed that the big cat had moved to Thrupp. Coryn Memory was one of several villagers who had seen it near their homes. She said it was about the size of a collie with a very long tail. Frank Tunbridge joined her in the search.

23 January 2010 Source: *The Stroud News and Journal*
They published pictures of pawprints in snow, found near Thrupp, that might have been from a large cat.

2 February 2010. Source BBC Website:
Frank Tunbridge commented on CCTV footage of what appears to be a jungle-cat in Stroud. The footage and commentary is available on the BBC news website at http://news.bbc.co.uk/local/gloucestershire/hi/people_and_places/nature/newsid_8491000/8491497.stm

16 February 2010. Source: *The Wiltshire and Gloucestershire Standard*
At five to one on the previous morning a woman and her partner were driving home from work when they spotted a big black cat on Cranhams Lane in Chesterton.

On 2 September 2009. Source: *The Evesham Observer*

There was an art expedition in Pershore. The organizer, Jess Tomlinson, used materials to recreate old Worcestershire folktales. One of her favourites was the story of the black cat of Broughton, which inspired several pieces in the exhibition. This cat was said to roam the countryside but had never been captured.

That is an apt place to end the chronology of newspaper reports and, indeed the section on big cats. Worcestershire and Gloucestershire folklore is full of stories about animals. Some of these will be discussed in the following chapter.

The Scottish wildcat is the only wild felid that is supposed to live in the British Isles

The Black Dog of Bungay, title page of the famous 1577 pamphlet about the Black Dog of Bungay, public domain, downloaded from http://www.bungay-suffolk.co.uk/images/black-dog.gif, 15 May 2010.

FOUR

Paranormal Portents

There are those who believe that the science of Cryptozoology is undermined by diversions into aspects of the paranormal. I justify the excursion in this chapter by referring to the book's title. Mystery Animals should, in my view, include those creatures which are not now regarded as sightings of flesh and blood animals.

In 2006 Merrily Harpur wrote a book, *Mystery Big Cats* in which she suggested that the cats could be modern day spectres or phantoms and linked to the black dog phenomena. Earlier another big cat researcher Di Francis had argued that the black dog sightings could have been of big cats.[26] Given that many sightings of black animals occur at a distance, often from a car on unlit roads, confusion between felid and canid in some cases could be understandable.

The black dog has donated its name to a public house in Newent and to a road, Black Dog Way, in Gloucester. Usually bigger than normal dogs, with glowing eyes, black dogs are said to appear without warning at places such as cross-roads, places of execution, old paths and after electrical storms. Amongst the most famous are the Black Dog of Bungay in Suffolk which allegedly left scorch marks on the church door and the Black Dog of Newgate, which haunted the notorious London prison. Localised names such as Black Shuck in Norfolk hint at the danger they were thought to present as they haunted dark lanes and terrorised innocent travellers.

The connotations afforded black dogs are widespread. Winston Churchill used the term black dog to describe his depression but he was not the first to do so.[27] Hester Thrale, Samuel Johnson and James Boswell all made the connection in their eighteenth century correspondence. Sir Walter Scott did the same in 1826 and Charles Dickens noted that, in childhood, he was scared of tales of the black dog.[28] This fear perhaps stems from mediaeval beliefs that the black dog was a familiar of witches and, sometimes, Satan himself. In 1127 the Anglo Saxon Chronicle reported that hideous horsemen in groups of twenty to thirty had travelled through woodland in Lincolnshire and Cambridgeshire. They rode on black horses and goats and were accompanied by black hounds.

The Wild Hunt, known in many parts of Europe, was a collection of ghostly huntsmen, carrying the tools of hunting, who ran through the skies or along the ground. It may have been invented by the ancients to account for thunderstorms. A man called Callow was said to lead the hunt at Feckenham in Worcestershire. He also appeared near Tenbury where his grave is marked and hideous black dogs have been seen running round the spot. Black dogs are also said to appear at Callow's Leap at Alfrick where Callow jumped down a precipice.[29]

Another reported leader of the wild hunt was Sir Peter Corbet, who died in 1300. In 1281 he was commissioned by Edward 1 to hunt wolves in the royal forests of Gloucestershire, Herefordshire, Worcestershire, Staffordshire and Shropshire. He owned Chaddersley Corbet and some land around Alcester. A legend says that he heard about his daughter planning to meet a lover in the woods and locked her in her room before releasing his hounds into the same woods. The lover was torn to pieces and the daughter drowned herself in a moat. Sir Peter hanged his hounds and threw their bodies in a pool. As punishment he was doomed to forever roam the forests with the hounds.

A further phantom horseman was Harry-ca-Nab who kept his dogs at Halesowen which is now over the Worcestershire border in the West Midlands. On stormy nights he would hunt boar or wild bull across the Lickey Hills. Those who saw his hunt suffered bad luck and even death.

Sightings of the black dog were often thought to indicate an imminent death. There are several examples in European mythology where the underworld is guarded by large and hostile dogs. This may be due to the scavenging nature of dogs and other canids. In the days when burials were more primitive such creatures would feed on dead humans.

Black Dogs have been reported on several occasions in Gloucestershire and Worcestershire. A summary of some of these sightings, listed alphabetically by location, follows.

Bewdley
In 1943 Stephen Jenkins saw a black dog trot past him.[30]

Bisley
One afternoon in December 2003 a couple were leaving the church when they encountered a large dog with a black coat and dark red eyes. This made the motions of barking at them but no sound was heard. They retreated into the church where it did not follow.

On 25 October 2006 Tristan Swayles referred to the Bisley sighting in an article on Black Dogs in Gloucester which was posted on the BBC's website. This prompted comments from readers, including two who claimed to have seen a black dog themselves.[31]

Bordesley Abbey
In 1864 James Woodward and his friend saw a black dog in the ruined abbey on two consecutive nights.[32]

Bredon
A black dog appeared in the bedroom of a cottage in Bredon during World War Two. It had glowing red eyes and warm breath. The witness was a four year old girl who had been evacuated.[33]

Jackament's Bottom
This is a farm between Kemble and Badminton where in 1835 a phantom black dog is said to have caused a coach crash. The driver apparently died when his horses bolted.[34]

Mickleton
Phantom black dogs have been seen in the area around Meon Hill, an Iron Age Hill fort said to be the inspiration of Tolkien's Weathertop in the Lord of the Rings. It is situated near Mickleton, Gloucestershire's most northerly point. On Valentine's Day in 1945 Charles Walton was found murdered by a farm known as the Firs. The murder appeared to be ritualistic and Detective Superintendent Robert Fabian from Scotland Yard was made aware of a 1930 book, *Folklore, Old Customs and Superstitions in Shakespeare Land* written by the Reverend James Harvey Bloom which related how in 1885 a young plough boy named Charles Walton saw a black dog several nights in succession. On the last occasion it was accompanied by a headless woman and, on that night, Walton's sister died. There is no evidence that this was the same Charles Walton who died in 1945. Following that murder a black dog was allegedly seen hanging from a tree close to the scene and it was rumoured that Fabian himself saw a black dog on Meon Hill. The murder was never solved and is known to crime enthusiasts as the Witchcraft Murder.

Minchinhampton
It was reported in 1912 that a woman from Minchinhampton was accompanied by a dog that she could see through.[35] The stretch of road adjacent to Hollybush Farm between Minchinhampton and the hamlet of Hampton Fields is known as Woefuldane Bottom, and whilst passing through on their way to and from Gloucester Market, carters were said to blindfold their horses in fear of a phantom black dog. In the 1950s a silent figure was seen following people in the gloom at this location. [36]

In 1976 Princess Anne purchased the Gatcombe estate which lies in close proximity to the hamlet of Hampton Fields in Minchinhampton. A local man Joe Hattersall told the Daily Express at the time that the entrance to the estate was haunted by a headless dog: He said he had seen it four times. Another local witness was Fred Webb, a retired builder from Minchinhampton, who claimed to have seen the headless dog whilst in his car.[37]

One website claims that this dog is known as the hound of Odin.[38] There were actually two hounds of the Norse God Odin. They kept watch outside the fortress hall, Lyfjaberg, one by day and one by night.

St Briavel's

A black dog is said to haunt the area between the castle of St Briavel's Great Hall and chapel. During a visit to the castle in March 2006 by members of the three counties paranormal group and a similar group, the Cheltenham based, Parasoc, a black lurcher was seen by a researcher in the corridor next to the dining area. A similar dog was also witnessed earlier in the evening by another member of the group in the kitchen area. Staff at the castle claimed that they did not possess a dog of this appearance.[39]

Sherbourne

The burial mound in the grounds of Lodge Park is said to be haunted by a black dog. A phantom coach and six horses driven by John "Crump" Dutton a seventeenth century squire has been seen in the grounds of Sherbourne lodge but not accompanied by the black dog.[40]

Standish

In a letter to the *Gloucestershire Citizen* on 2 November 2009 a lady from Swindon asked for verification of a legend connected to Park Farm, Standish. The story said that whenever anyone who had connections with the property died a black dog appeared on the lawn in front of the house. The letter writer said that the last confirmed sighting was in the late 1970s when the property owner said that the dog had started going inside the house.

Sudeley Castle

Sudeley Castle was used as a base for Prince Rupert during the English Civil War. According to legend his favour hunting dog was killed in a siege and now appears in the ruins of the banqueting hall shortly before a major misfortune strikes the castle's owner. Whilst this is the ghost of a normal dog, rather than a paranormal black dog, it is included to show how some of the folklore surrounding black dogs could also apply to other canids.

Swanpool Wood

A black dog occasionally appears from a limekiln in Swanpool wood, runs around Swan pool and returns to cover.[41] There is a possible link to the death of a woman and child.

Woodchester

In February 2004 and September 2005 a black dog was seen inside the Woodchester mansion, coinciding with the deaths of individuals associated with the trust. Earlier sightings between 1853 and 1966 were reported. There are several other stories of paranormal activity in the mansion, some of which are summarised on its website.

Winchombe

In 1892 it was reported that a gentleman walking late one night at Winchombe saw a man and a black dog walk through a closed gate.[42] There are similarities in many of the above tales and those reported elsewhere. Studies of black dogs have been done by noted folklorists such as Ethel Rudkin and Theo Brown. Brown divided them into three types:

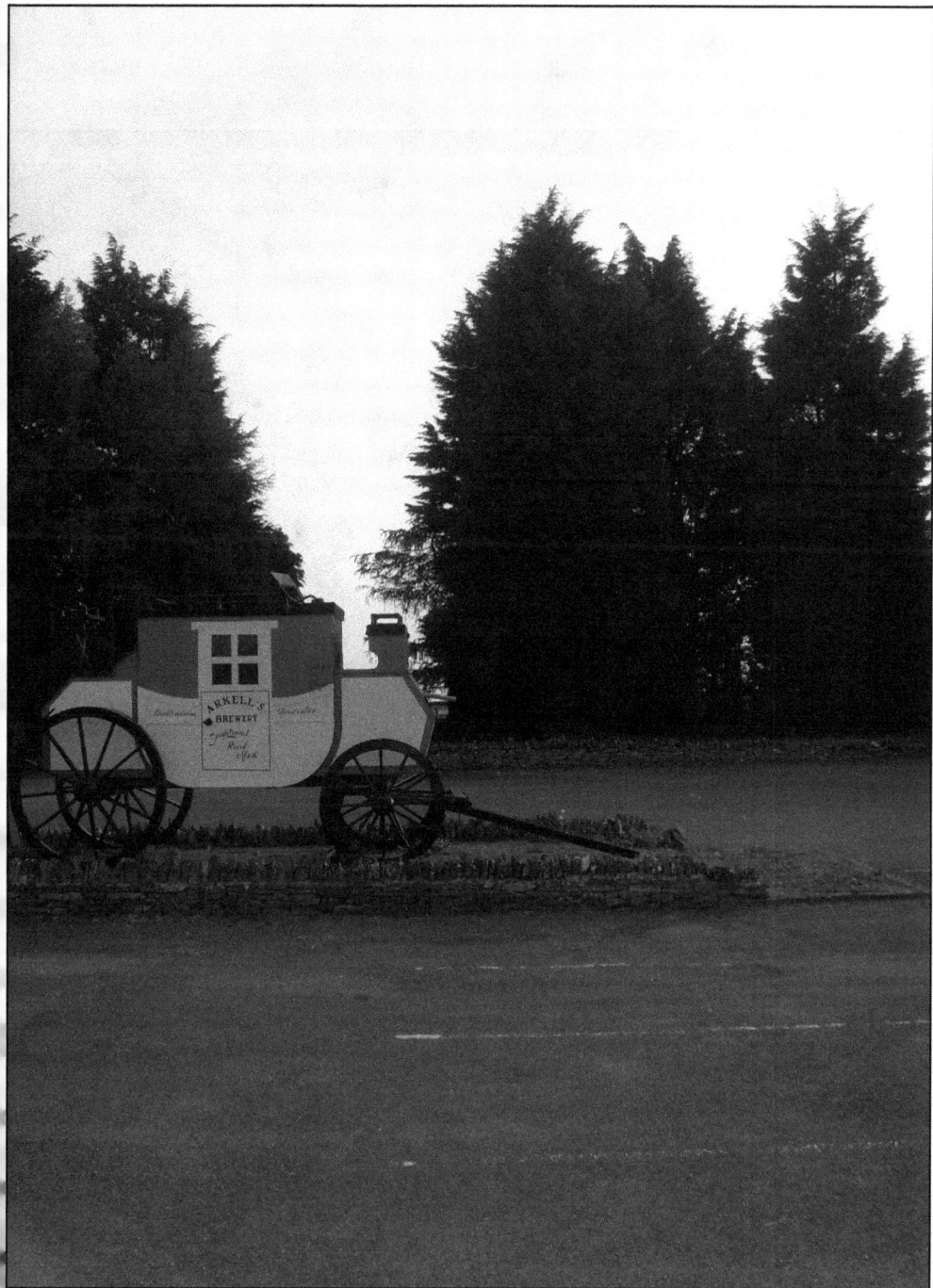

Model of an old-fashioned coach outside the Highway Man Inn at Syde, where a man saw a phantom coach crossing the road. Photo taken by author on 22 May 2010.

Sudeley Castle, supposedly haunted by a phantom dog.
Photo taken by author on 22 May 2010.

- A shape-shifting demon dog.
- A dark dog the size of a calf.
- A dog that appears in time with certain ancient festivals in specific areas of the country.

Katherine Briggs refined this classification further, to include those dogs that were the ghosts of men and the ones who were the ghosts of dogs. Of course it is not just men whose spirits were thought to return as a black dog. Eleanor Cobham, Duchess of Gloucester, made famous by Shakespeare was accused of trying to assassinate King Henry VI using witchcraft. This resulted in her being jailed for life after doing public penance. A black dog representing her spirit is believed to haunt Leeds Castle in Kent and Peel Castle on the Isle of Man.[43] She was imprisoned in both places.

Other animal spectres have been seen in Gloucestershire and Worcestershire. At Kingswood House, opposite Lady Katherine Berkeley's school at Wotton-under-Edge a horse drawn carriage is said to enter through one of the gates, drive around the house and leave by the other gate. Different phantom coaches are sometimes seen on stretches of road at Wickwar, Saddlewood and Frocester Hill. In 1972 a Mr Jeffries was driving a delivery van towards Cheltenham on the A417 at 17:30. He had just passed the *Highwayman Inn* and was close to the Syde turning when a white coach drawn by four horses crossed the road in front of him. A woman driver and passenger from Charlton Kings have twice seen a coach and four horses emerging from a track on the A40 by the *Puesdown Inn* and crossing the road. A short distance away a phantom horse and trap near the Hangman's Stone on the Salt Way have caused accidents. The same is said of a carriage and four horses on the Bream Road leading out of Coleford.

Stepping Stone Lane, the old coach route from Painswick to Stroud, takes its name from a mounting block close to where a footpath joins the road. Between this point and a milepost noting that is a further seven miles to Gloster, more black dogs have been seen. A goblin like figure is said to dance on the milestone.[44] A goblin is also rumoured to haunt the woods at Mickleton. He was known as the Mickleton Hooter, or Belhowja, and had his home in Mickleton Hollow. His howling and shrieking was often heard. Some people attributed this to a noisy vixen.[45]

Sometime around 1802 a servant was walking past a crossroad called Tinkers Cross in Leigh when he saw an animal resembling a lion with eyes as big as saucers. When reporting this in 1852 Jabez Allies quoted an old tract on goblins which described Puck as having saucer like eyes.[46] Had the sighting occurred in more recent times it might have been cited as one involving a big cat.

Not all ghostly animals were considered to be malevolent. One story relates that King John was hunting at Staunton when one of his courtiers displeased him and was sentenced to death. A reprieve was granted on condition that a suitable buck was found. One promptly appeared.[47] Other folktales tell of equally helpful creatures. The Mabinogion is a collection of Welsh legends, with the first five tales relating to Arthurian lore. One of these relates that a talking salmon sensed evil on his approach to Gloucester. He took two of King Arthur's knights, Kay and Gwrhyr, back up the river where they heard crying from within Gloucester castle. It was

Mabon, son of Modron, who had been kidnapped when he was just three days old. Arthur immediately marched to Gloucester and attacked the castle. Meanwhile Kay and Peredur went up the river on the back of the salmon and rescued Mabon who later became one Arthur's knights.[48] The Mabinogion also tells how the Twrch Trwyth, a king, was punished by God for his sins by being turned into a wild boar. He raided the Severn Valley but was eventually driven out to Cornwall by Arthur's Knights.

Another folktale tells of an encounter between a fisherman and a huge salmon to explain why coracle fishing ended on the lower middle reaches of the Severn.[49] The man netted the fish and hauled it abroad but his coracle sank with the weight. The use of folktales to explain events is also seen in the following story which accounts for the presence of a monolith, the tibblestone, by the roadside at Teddington Hands. A giant who lived on Dixon Hill, a few miles away, used to throw stones at ships. Once he slipped and the stone landed at Teddington instead.[50] A similar story is told about the Devil who sat on Meon Hill and was annoyed by the sight of Evesham Abbey. He kicked a huge story at the abbey but it swerved to the right.[51]

Throwing stones at Christians was a popular pastime for Satan. A limestone rock formation in the shape of a crooked chimney in a quarry in Leckhampton is known as the Devil's Chimney. The Devil was said to have sat on Leckhampton Hill and hurled stones at Sunday churchgoers. They turned the stones back and forced him beneath the ground where he began using the stones as a chimney to unleash the fires of hell.

The Roman iron workings at Bream are called the Devil's Churchyard. One day at dusk a man gathering firewood saw a figure with horns and a cloven hoof climbing down a tree.[52] Perry Wood in Worcestershire is said to be the place where Oliver Cromwell made a pact with the Devil.

Many animals were connected to the Devil. The medieval chronicler William of Malmesbury, writing in the 1120s, tells how, in 1065, a witch living at Berkeley was told by her pet jackdaw that death was approaching. She asked her family to protect her body by sewing it inside a stag's hide and placing it in a stone coffin fastened with three chains which had to stand inside the church for three nights. On both the first and second nights a chain was broken by demons then the Devil seized her and carried her away on a black horse.[53]

Our ancestors believed that they shared their environment with all manner of strange creatures. They often depicted these in art and sculptures and many examples survive in their churches. Dragons are commonly seen, for example in the external sculptures at Fairford Church where a wyvern and griffin also appear. Half of the forty gargoyles at St Peter's church in Winchcombe represent dragons.[54] Others include representations of local people as well as the typical devils and monsters.

Protruding from the west wall at Deerhurst church are two animal heads, which gave rise to the local tradition of a dragon. According to Sir Robert Atkykn's book, *The Ancient and Present State of Gloucestershire*, first published in 1712, a serpent was poisoning the inhabitants and killing their cattle. The inhabitants petitioned the king who promised an estate on Walton

Hill to whoever killed the beast. A labourer named John Smith took up the challenge. He set out a trough of milk. After drinking this the dragon went to sleep and ruffled its scales. John took an axe and beheaded it. In the late eighteenth century a Mr Lane who had married into the Smith family claimed to still have the axe. A century later the Revered George Butterworth found that his parishioners still spoke of the flying adder. In 1991 a farmer in the parish, called Smith, had a hillock on his land called the Dragon's Trump.

The belief in dragons is attested to beyond the influence of religion. A hill in Newland was called Drakehord, the place of the dragon's treasure in 1337. A map of 1839 shows it as Dragons Ford. There was a standing stone called the Dragon's stoke or drakestone near Stinchcombe. There was also a place called Drakestone near Sherbourne. Drakelow in Wolverley was supposed inhabited by a dragon in 1582.

When the porch at St George's church in Cam was restored in the mid nineteenth century a sculptured boss showing the eponymous saint and his dragon was set in the vaulting. Moreton Valence's north door tympanum shows St Michael spearing a water-monster. St Michael's quarry depicted on the Norman tympanum at Harnhill Parish Church is a dragon. Like George Michael was a patron saint of chivalry. He is associated with the dragon because the book of Apocalypse, also known as Revelation records that *"there was a great battle in heaven, Michael and his angels fought with the dragon."* [55]The dragon represented Satan and many artists copied this connotation.

A thirteenth century window at St Mary's Kempley depicts the struggle between St Margaret and the dragon. This involved the dragon swallowing the saint but she escaped alive when the cross she was carrying irritated the dragon's innards and forced him to release her. The message here is that anyone can be rescued from the devil. Margaret of Antioch was a popular saint with over two thousand churches dedicated to her, more than two hundred and fifty of them being in England.

Twin dragons are depicted on the clerk's desk at Stoke Bliss near Tenbury which say Roger Osland Churchwarden 1635. Other dragons look down from the nave roof at Eckington and appear on the fort at Elmley Castle. Given that dragons were associated with the Devil and thus were enemies of Christ it may appear strange that they are depicted in churches outside of images of battle. The reason is probably symbolic as the images showed that they had been conquered, and tamed by the Church.

The outline of two beasts is cut into the walls of North Cerney church. One, four feet tall, has the body and legs of a dog, the hooves of a pony, a thick tail and the head and arms of a man. The other is more like a leopard. They may be representations of the manticore, a beast described in a mediaeval bestiary as follows:

> "It has a triple row of teeth, the face of a man and grey eyes; it is blood-red in colour and has a lion's body, a pointed tail with a sting like that of a scorpion and a hissing voice. It delights in eating human flesh. Its feet are very powerful and it can jump so high that neither the largest of ditches or the

broadest of obstacles can keep it in".[56]

At Fairford St Mary's there are carvings of four dwarves which stand guard on the church tower. At Lechdale the Church of St Lawrence has a monster on the outside of the tower.

A Norman tympanum at Dumbleton shows the head of a man with the ears of an ass and three pieces of foliage coming out of his mouth. This motif occurs in wood or stone at other churches including Cam, Compton Abdale, Guiting Power, Lechdale and Quenington. Green men are carved in Gloucester Cathedral.

The sculptured toad in Berkeley Church represents a real toad which supposedly grew to a monstrous size by devouring the bodies of prisoners who died in Berkeley castle. A carving of the toad is in the morning room at the castle. John Smyth, steward to the Berkeley family, in the 1600s said that the toad lived in the reign of Henry VII and had lived for several hundred years. In 1836 Grantley Fitzhardinge Berkeley wrote about the castle and made the following comment on the toad:

> "One thing, which I remember well as a child has been removed; on yonder shelf was the stuffed skin of a huge seal, often pointed out to me by my nurse as the great toad of old, which my ancestry used to keep in the donjon to feed upon their captives; and to which, as an ancient legend run, (doubtlessly derived from as authentic a source) the Marquis of Berkeley was supposed to have abandoned his two children."[57]

A misericord is a small wooden shelf underneath folding seats in churches. They often contain interesting depictions of secular or pagan images and scenes that contradict their Christian setting. The misericords at Fairford include a Green Man and two wyverns. Those in Gloucester Cathedral show St Michael slaying a dragon, numerous monsters, a mermaid and a wyvern. The misericords at Worcester Cathedral include a knight fighting two griffins, a lion and a dragon fighting, a sphinx, a dragon and a cockatrice. The priory church at Great Malvern has misericords which include a mermaid, a cockatrice and a wyvern.

Other relics of past beliefs can be found in place names. Roy Palmer in his book on Gloucestershire folklore points out the many fairy place names such as Puckpit Lane between Hailes and Winchcombe, Puck's Well below Humblebee Wood at Sudeley, Puck Acre in Daglingworth, Puck Moor in Dymock, Puckpit Meadow in Aston Somerville, Pug's Path in Wotton-under-Edge, Puck Pit in Maisemore, Dobb's Hill in Eldersfield and Hob's Hole in Coppice Blockley.[58]

In the early twentieth century a Churchdown man named Enos Berry saw a fairy cortege on Chosen Hill and made a remark to them about being late for their funeral. They flew off into the air, apparently headless. Other people had the same experience.[59] An old man at Amberely saw fairies dancing and became known as Cold Water Jesse because he gave up drinking cider following the experience. In a common folktale motif a traveller found an inn at Dursley where he was welcomed but his hosts had gone in the morning. He left two guineas for his stay and returned to the spot to find the coins lying in the grass with no sign of the inn.

A woman drying her hair in the forest of Birdlip Beeches saw a fairy in the early twentieth century. He was described as being the colour of dead aspen leaves, nine inches high, misshapen, wrinkled and ugly. He complained after being caught in the woman's hair. She mentioned the experience to a professor at Bristol University who was not surprised and knew that fairies still lived in Birdlip Beeches.[60]

Gervase of Tilbury tells how in a Gloucestershire forest there was a glade in the midst where a hillock rose to the height of a man. Knights and hunters climbed the hillock to obtain relief from heat and thirst. A fairy cup-bearer would present a large drinking horn containing a wonderful liquor that cured the ailment. One day a knight from the City of Gloucester decided to keep the horn. He was sentenced to death by the Earl of Gloucester who presented the horn to his sovereign, King Henry I.

Death was the punishment for witchcraft through the middle ages. On the continent witches were often accused of committing crimes in the shape of a wolf. Due to the early extinction of wolves in Britain there are not many werewolf stories. Instead those witches thought to have the power to change themselves into animals were associated with more mundane beasts. A witch Susan Sly who was depicted in a mural at the *Bat and Ball Inn* in Churchdown once lived in Buttermilk lane. She used to change herself into a hare and was once bitten on the back leg as she fled from hounds in that guise. An old woman from the Edge Hills in the forest of Dean was less lucky. Caught by hounds in the shape of a hare her human body was found the following day with the wound.[61]

T A Ryder from Elton near Westbury-on-Severn heard a tale in which an old man guarded his daughter from suitors. One disguised himself as a hare to get close to her but was shot. When the old man saw the boy he walked with a limp.

There is some evidence that the animal in this type of tale may change to suit local circumstances. One story, supposedly told within twenty years of the event, relates how, in 1850 a huge black cat in Worcester Street Kidderminster scratched at the door of local witch Becky Swan. She opened the door, turned pale and let it in. Three days later her ashes were found in the house.[62] This is a story that could be told equally well with a black dog and brings us back to the theories alluded to by Francis and Harpur that the two were, and are, interchangeable. It is only in the last decade or so that people have been encouraged to report sightings of big cats. Any such sightings in previous centuries are unlikely to have been recorded and explanations were likely to be sought in spiritual rather than physical terms.

Like the wild boar, steam trains have returned to parts of the Forest of Dean. Photo taken by author on 31 May 2010.

FIVE

A Caiman in the canal and other cryptozoological chriosities.

This chapter looks at some of the other incongruous or unusual animals seen or discovered in Gloucestershire or Worcestershire. It is arranged in alphabetical order with unidentified animals listed either as "beast of" the place they are most associated with or by the name of that place.

Baboons

On 8 July 1974 eighty baboons escaped from the West Midlands Safari Park. Within days the majority were recaptured or dead. In September and October the four who survived were seen in Oldington Woods near Kidderminster.

Bears

The *Western Daily Mail* of 1 May 1889 carried the following story.

> "The Attack Upon Frenchmen in Dean Forest.
>
> Someone, if we remember rightly, it was Doctor Charles Mackay, wrote a book a few years since on the Madness of Crowds. If he brings out a second edition he will be able to add a chapter on the mad conduct of certain colliers and others in the Forest of Dean on Friday last. It is necessary briefly to recapitulate the facts and for this purpose nothing can be more effective than the plain simple narrative of the local reporter. From this it would seem that on the afternoon of the day named four Frenchmen with two dancing bears visited Cinderford and that, after parading the principal thoroughfares and performing several feats they left the town with the animals and proceeded quietly along by Nailbridge for Drybrook about half past six o'clock. By some means or other a rumour was circulated at the

former place that one of the bears had killed a child in Cinderford during the day. This simply infuriated the inhabitants in the neighbourhood who assembled in large numbers, armed with sticks, stones, pikes and forks. When they Frenchmen arrived at Nailbridge they met with a very warm reception, a crowd of people following them all the way to Ruardean, beating both the men and the bears at intervals in a most savage manner. On arriving at Ruarden the angry mob killed one of the bears and commenced a further attack on the Frenchmen who took to their heels and fled for their lives, leaving the live bear behind them. The poor animal made off as quickly as possible towards Bishop's Wood, followed by a large crowd, yelling and hooting more like wild Indians than Englishmen. At Bishop's Wood a tremendous onslaught was made on the poor animal and there in a field they beat it to death."

The report goes on to express sympathy for the Frenchmen, and announces the establishment of a fund to raise money for them. Two days later it reported that fourteen men had been prosecuted.

Beast of Ankerdine Hill

In 2003 a creature described as half fox cub and half wild boar was seen on Ankerdine Hill. The *Worcester Evening News* reported on 1 August 2003 that on 30 July that year the creature had walked out in front of a car at approximately 16:00. Jo Morris, from Suckley and her mother, Rachel were in the vehicle. Jo described the beast as being the size of a half-grown fox cub with a long nose, small ears and mottled brown skin. She likened it to a hyena and felt that the long thin tail precluded it from being a fox cub. Both witnesses dismissed the possibility that it was a coypu and said they had discussed the sighting with vets who were equally mystified. The newspaper published an artist's impression of the beast.

On 5 August 2003 the *Worcester Evening News* reported that Shelia Harris had seen a similar creature wandering in the road at Hadley Heath, near Droitwich on 5 April. She said it was larger than the beast described by Jo Morris and initially thought it was a hyena with a mane. More conventional theories put forward including a fox with mange and a Muntjac deer.

A lion headed maned hairless animal was killed on the Isle of Wight in 1940 and revealed as a fox in an advanced state of mange. Could similar creatures explain both the Beast of Ankerdine Hill and the Ashchurch lion?

Beast of Badminton

Some mystery animals have left signs of their presence without ever being identified. In November 1905 a mystery animal killed sheep in the area around Great Badminton, South Gloucestershire, leaving the flesh virtually untouched but lapping up the blood.

The Daily Mail on 1 November 1905 quoted Sergeant Carter from Gloucestershire Police as saying that this wasn't the work of a dog because they were not vampires.

The *Penny Illustrated Paper and Illustrated Times* published the following comment on 25 November 1905.

"Badminton whose sheepfolds are again being ravaged this week offers a field of investigation for the naturalist. A jackal, it is commonly reported, is the culprit. The tale of dismal howlings by which the work of destruction is said to be accompanied certainly does suggest this animal. The fact that the blood is drawn from the victim and the body left is against the jackal theory. Armed with terrific teeth, the jackal has nothing of the vampire in his procedure; bone and flesh are his mainstay. He crocks and eats bones which less formidable jaws would only gnaw. There should not be insuperable difficulties in the way of detecting the offender at Badminton. Every predatory creature leaves a mark which the naturalist reads. The stoat, the weasel, the ferret, the dog, the cat, - each has its distinct method. One creature seizes its victim in a particular place and opens a certain vein, another will skin its prey with the skill of a professional. The dog is known from his rough and tumble style of execution.

In such a case as that under consideration the naturalist is decidedly the best detective. From the signs which he reads he knows in an instant by what savage creature the sheepfolds of a countryside are being ravaged and can reconstruct the crimes of the night. The poultry farmer comes down some morning to find every fowl and chick lying dead in its pen. To him it is impossible to account for the outrage unless he can blame some wanton boy. The man with the seeing eye however can tell him that an animal is the culprit, a ferocious little savage which, when its appetite is appeased, kills, apparently, for the very lust of slaughter just as a madman kills when he runs amok. When, however, it is the sheep which are dying as now at Badminton, the matter is much more serious and, as the evidence shows, mysterious."

The report goes on to say that some years previously farmers dwelling in the vicinity of the Earl of Roden's estate were the victims of outrages similar to Badminton. Despite all the hysteria about vampires a couple of pine martens were identified as the culprits.

Also on 25 November 1905 the *Bristol Mercury* declared that the Badminton beast was a jackal which had escaped from a menagerie in Gloucester.

No records have been found of such an escape.

National interest waned when in December 1905 it was reported that a large dog had been killed near Hinton but there had been no reports of unusual sheep killings in that area.

We cannot now tell what the animal was but the accounts of a strange beast killing sheep will be horribly familiar to farmers who have lost livestock in similar circumstances more recently.

Beavers

On 28 October 2005 six European beavers were released on an enclosed site in South Cerne, Gloucestershire, the first time that this particular species had lived in the wild in England for 500 years.

Two years later they were successfully breeding. On 26 July 2008 the *Daily Mail* reported that the first kits had been born. The closely related species of Canadian Beavers have been introduced to various parts of England over the years. These include Ashdown Forest in East Sussex/Kent and on the River Axe in Somersett. An article in *Do or Die #8* reports that

> "the two most significant incidents so far have been in Ashdown Forest (East Sussex/Kent) and on the River Axe in Somerset".

Boar

Those who dismiss the concept of unknown animals being found in Gloucestershire or Worcestershire would do well to learn from the wild boar experience. Boar were exterminated in England during the middle ages.

A traditional Worcestershire Ballard tells the story of Sir Ryalas, who slew one in Bromsgrove.

> *Now bold Sir Rylas a-hunting went*
> *All along and down a lee*
> *Now bold Sir Rylas a-hunting went*
> *Down by the river-side*
> *Bold Sir Rylas a-hunting went*
> *To catch some game was his intent*
> *Down in the grove where the wild flowers grow*
> *And the green leaves fall all around*
>
> *He spied a wild-woman sitting in a tree.*
> *All along and down a lee*
> *"Good Lord, what brings you here?" said she.*
> *Down by the river-side*
> *"Oh there's a wild boar in the wood*
> *He'll eat your flesh and drink your blood"*
> *Down in the grove where the wild flowers grow*
> *And the green leaves fall all around*

This particular beast had killed a Lord and gored thirty other men. Real boar can be dangerous to people but not on that scale. It has been claimed that Bromsgrove was originally called Boar's Grove but, if so, this would have been before the Anglo-Saxon charters and Doomsday Book.[63]

In 1802 there were several reports of a large boar with an unearthly roar which was crushing fences and hedges in the Forest of Dean. This became known as the, original, Beast of Dean but a search by villagers from Parkend failed to find it.

Wild Boar started to reappear in Britain in the 1980s, mainly in the form of escapes from farms which were breeding them for meat. In 1988 the government confirmed the presence of two boar colonies in Dorset and on the borders of Kent and Sussex. On 25 February 2004 a written answer in the House of Lords was quite specific that there were only three locations where wild boar lived in Britain. These were Kent/East Sussex with between one hundred and two hundred animals, West Dorset with between twenty and thirty animals and Herefordshire with between ten and twelve animals.[64] In 2008 a government report confirmed that these were all breeding colonies and recognized an additional site in the Forest of Dean. This came after wild boar had been filmed in the forest for the BBC's *Autumwatch* programme, broadcast in November 2007, and a wild boar had been shot by a park ranger when it ventured near a primary school in January 2008.

Prior to this official recognition wild boar had been seen by many witnesses in and around the Forest. They had returned to an area where, in 1254, they had been so numerous that a hundred could be taken for a Christmas feast.[65] On 5 December 2002 a wild boar was hit and killed by a car, driven by Richard Kevill, on the A40 between Gloucester and the Forest of Dean. A spokeswoman for DEFRA said that they had no knowledge of wild boars in Gloucestershire and were trying to trace the owner. The only licensed boar owner in the region was based in Wiltshire.[66]

Another boar was sighted in Brockweir in July 2002 twice in consecutive days.[67] On 9 October 2003 The *Birmingham Evening Mail* reported that wild boar were living in woodland on the Herefordshire and Worcestershire borders. The Gloucestershire sightings were explained as strays from a population in Ross on Wye or possibly released from elsewhere. In January 2004 one of the creatures, which can weigh up to two hundred kilograms, found its way into a supermarket into Cinderford and knocked over an elderly female shopper. After being confronted by the shop's staff it ran off into woods by Bellevue road.

The Gloucestershire Citizen reported on 29 November 2004 that wild boar had invaded the village of Staunton, roaming the woods and lanes and digging up a lawn. Resident Jack Cockburn found six of them rooting in his flower beds. There were two mature females, two half-grown ones and two very young. A spokesperson for the Forestry Commission said that the boar were herded back into the woods and seemed very tame.

In November 2004 a wild boar sighting in Coleford, involving rooting damage was reported to DEFRA. They confirmed the report, noting it as a recent escape or release.[68] The same month boars chased horsewoman Carla Edmonds through woodland on the Gloucestershire/Monmouthshire border. The creatures later added insult to injury by digging up around one hundred square feet of one of her fields. Despite this apparent vendetta, Ms Edmonds described seeing the creatures as "amazing" in an article on the BBC's website on 29 November 2004.

On 22 March 2005 the *Birmingham Post* reported that the Forestry Commission had decided to erect a series of four-foot high mesh fences around campsites in the Berry Hill area of the forest in order to protect tourists from the boar. This was three years before the Government officially accepted that boar resided in the forest. Remember that it took the Forestry Commission eight years to tell the public about big cat sightings in the Forest.

In October 2006 a wild boar in Broadway, Worcestershire, was reported to DEFRA who felt that the evidence was not conclusive.[69] The following month a boar in Suckley, Cradley, Worcestershire was reported as causing injury to a dog. This report was also considered to be inconclusive by DEFRA.

On 24 January 2008 the *Gloucestershire Echo* reported that wild boar were continuing to cause misery for some people. Rose and Terry Baldwin had taken care of common land between their garden in Brierley and the Forest of Dean for thirty years. Then the boar began raiding it every night, ploughing up the grass in their quest for food. On the other side of the Forest a golden retriever was gored by a boar on Blakeney Hill.

On 31 January 2008 a wild boar broke into the grounds of Ruardean Primary School. When it became aggressive a forest ranger shot it. Defending the decision against harsh criticism a spokesperson for the Forest Commission was quoted in the Daily Mail as saying that they had found dogs ripped apart in the forest and had received several reports of the boar being aggressive.

On 20 February 2008 the *Gloucestershire Citizen* quoted DEFRA as saying that landowners with a firearms licence were entitled to shoot boar on their property. In 2004 Lord Swinfen had asked the House of Lords with what weapons it was legal to kill wild boar. Lord Whitty replied that:

> "there are no specific legal restrictions governing the use of weapons to kill wild boar but the use of bows, crossbows and any explosives other than ammunition for a firearm to kill any wild animal is prohibited."[70]

In November 2008 Soudley Football Club was forced to cancel a home game against Charfield because wild boar had munched their way across the pitch. This can be put into perspective by the ten thousand wild boar that live in Berlin and once damaged the pitch of Hertha Berlin, the city's largest football team.[71] On 7 July 2008 the *Gloucestershire Citizen* reported on its front page that Russell Baker from Bradley Hill, near Soudley, had built a wall around his property after boar tore up his lawn.

Damage wrought by boar is visible whilst big cats leave very few marks. For this reason it is harder to deny the existence of the boar and their success in re-establishing themselves in the forest bodes well for all wildlife, despite the opposition of some local people.

Caiman

Boar might have been reluctantly accepted but few people seriously entertain the possibility that a caiman, a crocodile like creature, could live in the Gloucester-Sharpness canal. This sixteen mile long waterway links the two docks. It is sixteen feet deep and, at one time, was the broadest and deepest canal in the world.

In 2003 a sighting of an unusual animal in the canal caused a media frenzy. Richard Lacy, an experienced bridge keeper, was on duty at Sellars Bridge: a narrow swing bridge adjacent to the *Pilot Inn* at Hardwicke. At approximately 09:55 one morning in June 2003 he claimed to have witnessed a duck make a sudden takeoff from the canal, narrowly escaping what appeared to be a three foot long reptile with legs. *The Sun* reported this story on 20 June 2003, the day after it published a report of a crocodile in Cannock, Staffordshire[**]. Six years earlier it had stated that a three-foot reptile, believed to be a small crocodile had been spotted in Regent's canal in Hackney.[72]

There had been rumours of a crocodile or caiman in the Gloucester Sharpness canal before Mr Lacey's sighting. In addition Charles Fort refers to four incidents of crocodiles being killed or seen near Over-Norton in Oxfordshire between 1836 and 1862.[73] Mary Chipperfield in her 1972 book Lion Country stated that she kept a book of wild animals seen in case they were connected to the circus that she owned. These included a ferocious animal with snapping jaws seen in Gloucester but never captured.

Caiman, and similar species, are not hard to obtain in Britain. The RSPCA rescued more than forty alligators and crocodiles between 2000 and 2005. In 2004 a man from Edinburgh, Anthony Quinn, was arrested after police discovered a spectacled caiman, in the boot of his car. He had brought it on the internet for £250 and kept it in a bath in his flat. Officers responded to an advert offering the caiman for sale. Two years earlier two spectacled caimans were retrieved from a council flat in Oldham.

Another keeper working on the Gloucester and Sharpness canal told a researcher representing the CFZ that he had seen a rabbit killed by a polecat and that terrapins were also there. The researcher, Andy Stephens, had received a separate report of a polecat in the area. The zander, *Sander lucioperca*, is in the canal and may weight fourteen pounds. This species reached Britain from Eastern Europe in 1963. Although often described as a pike and perch hybrid they are actually a separate species.

The large pike in the canal have eyes towards the top of their heads, resembling a crocodilian. Similarly, the pelvic and pectoral fins on a large pike might be mistaken for legs during a brief glimpse of a sighting. Mr Lacey felt that the animal seen was not a pike. He had been working on the canal for forty years and had been fishing for forty eight years. Interestingly he explained the previous sightings of a caiman as being terrapins which are not indigenous.

A columnist for the Worcester news, John Philpott, stated in his column on 7 February 2008 that he had discovered the jawbone of a large pike whilst walking along the River Severn on Sunday 1 February 2008. The bone measured more than five inches in length and had several

---

** CFZ researchers investigated both incidents Andy Stephens' report from 2003 can be found in Appendix One and online at http://www.cfz.org.uk/expeditions/03croc/gcroc.htm

teeth, some of which were about half an inch long. He identified the fish as *Esok lucius*. One of this species, caught by John Murray in 1798 and measuring more than six inches was said to have weighed seventy two pounds. Mr Philpott calculated that his specimen would have weighed sixty pounds, more than twice the usual weight of pike.

November 2005 saw the publication of a report on canal wildlife diversity by British Waterways.[74] This included a report of a crocodile near St Cyr's church on the Stroudwater canal, Stonehouse. This may have been a confusion with Hardwicke. The BBC quoted a British Waterways spokesperson as saying that it was unconfirmed. Other animals sighted in canals nationally according to the report included mink, seals and dolphins. On 21 November 2007 Mark Robinson, national ecology manager, at British Waterways was quoted in the Independent as saying that the Stonehouse crocodile was probably just a large pike.

The creature in the Gloucestershire Sharpness canal might have been an abnormally large pike but the possibility that it was a caiman or even a crocodile cannot be completely discounted. On 20 July 2001 the Independent reported that the fire brigade in Vienna had pulled a crocodile measuring twenty four to twenty eight inches from the River Danube canal. According to the BBC's report a crocodile had escaped into the River Rhine in Germany earlier that month but it was much longer in length.

Capybara

The Capybara is the largest rodent in the world and can grow to the size of a sheep. Originally from South America they are not indigenous to Britain. On 13 July 2001 the *Daily Express* reported that some had been spotted in the Worcestershire village of Brotheridge Green. John Hodson of the Worcestershire wildlife trust believed that they had escaped from a zoo. One had been found dead in Tewkesbury and there had been other sightings, including three in Upton-upon-Severn which were noted by DEFRA in 2002.

Coypu

On 1 September 1998 a man telephoned police from the *Russell Arms* public house to report a sighting of a rodent like creature, ten times bigger than a rat, near the Hales Road in Cheltenham. He described it as making a high pitching squealing noise. Terry Hooper of the Bristol-based Exotic Animals Register considered that the animal may have been a coypu. Coypu were brought to Britain for their fur in the 1929 and some escaped although many were rounded up and exterminated in 1989.

In 2001 a man in a boat on the river Severn filmed a Coypu with his video camera. This followed a series of reports at various points between Tewkesbury and Upton-upon- Severn. One of these, in Upton, was reported to DEFRA who noted that the sighting "*confirms report of species.*"[75] The agency was more dismissive of a sighting in Apperley, Gloucestershire in October 2007, attributing it to the ambiguous "*other causes.*"

Eagle Man

On 2 January 2008 a poster using the pseudonym Alien Embryo contributed a report of an unusual creature to the unexplained mysteries website. His brother's friend had been walking in the Malvern Hills with his wife and another couple. The other man confronted a hunched vulture like creature that then followed the group. He then visited a spiritual person who claimed that the creature was a gatekeeper.[76] Given that the story is third-hand, with no supporting evidence, it cannot be substantiated.

Fish

On 28 May 1881, a thunderstorm in Worcester apparently brought down tons of periwinkles and hermit crabs.[77] The conventional explanation was that a fisherman and his assistants were responsible, despite not being seen and having no motive. The area covered stretched for about a mile, spreading from Cromer Gardens Road. Specimens were sent to the editor of Land and Water.[78]

Frogs

On 24 October 1987 it inexplicably rained frogs in Stroud. Hundreds of thumbnail-sized, rose-colored frogs fell from the sky, bouncing off umbrellas and pavements and disrupting people going about their business. *The Daily Mirror* and the *Daily Star* referred to some of the frogs being pink, relying on the testimony of an old lady. Both newspapers reported the opinion of naturalist, Ian Darling, who thought that the frogs were albino and the pink was caused by their blood vessels being seen through the skin. There was a theory that a whirlwind had carried the frogs from the Sahara to Gloucestershire. There is a precedent for this.

In July 1968 the Midlands were covered in what appeared to be red rain but was actually sand blown over from the Sahara inside a massive pressure system. Frogs also fell on Cirencester in 1986 and supposedly on Fairford in 1660.[79]

On January 14 1915 the *Dean Forest Mercury* reported that a small dark frog, with a yellow band on its back, fell out of some coal that have been in place for several thousand years and hopped away. This was in the Trafalgar colliery, the first in Great Britain to use electricity underground. Sir Francis Brian, the proprietor of the colliery wrote to the newspaper, saying that he proposed to put the frog in the British museum.

Iguana

A four foot long iguana was found in the Tesco car park at Quedgley in April 1995. It was taken to a wildlife rescue centre.[80]

Lizard

According to Stroud District Council a three foot monitor lizard escaped from a pet shop on 2 August 2004. It was found in the local area on 19 October 2004.[81]

Otters

Otters were virtually extinct in Worcestershire in 1984, due to pollution and lost of habitat, but have begun reappearing in many of the county's waterways.

They are sometimes killed by road traffic. Sir Christopher Lever noted six UK records of the Oriental short-clawed otter (*Anoyx cinerea*) between 1983 and 1993, including one in

Gloucester in November 1985. Footnote reference. *

Pine Marten **
The pine marten is a member of the mustelid family, about the size of a domestic cat. It was thought that they became extinct in the Midlands during the nineteenth century. This assumption was doubted when a dead pine martin was found in Chaddesley Wood on 15 June 2002. It may have been an escapee from a private collection but a lady had reported seeing a pine marten in the same woods in 2000. [82] In the wild pine martens have a lifespan of eight to ten years.

Porpoise
The *Gloucestershire Citizen* of 11 June 2004 reported that a porpoise had been discovered in the River Severn near Elmore lane in Stonebench. The mammal, which is the member of the dolphin family was seen by a pedestrian who spotted its fin going in and out of the water. The porpoise normally lives in salt water but can survive in fresh water. The RSPCA were called but decided not to move the creature as it was not showing any signs of distress.

Racoons
In January 2007 two young racoons escaped from the ark Animal Sanctuary near Evesham. The pair named Bonnie and Clyde, were not considered to be dangerous.

Seal
On 29 June 2006 *The Gloucestershire Citizen* reported that three days after seeing a big cat Emrys Davies claimed to have seen a seal by the old Wye Bridge between Tutshill and Chepstow. A seal was also reported to the Gloucestershire Naturalists Society in 1997.

Skunks
On 23 August 2009 the BBC's website reported the existence of a group of wild skunks in Coleford. One, named Ozzy, was handed into the Vale Wildlife Hospital in Tewkesbury. The founder of the centre was quoted in the Daily Express on 31 July 2009 as saying that there could be a family of wild skunks living in the countryside. Lyndon Booth photographed a skunk in the garden of a house in Tufthorn Close and saw another one.[83]

---

* *The Naturalized Animals of Britain and Ireland* (New Holland, London, 2009)

** It was thought to have become extinct over most of England in the latter years of the 19th Century, but in October 2010, Natural England confirmed that this was a mistake. As the BBC website confirmed:

> The pine marten, one of Britain's rarest and most elusive mammals, is back - and the reason is it never quite went away. A new report reveals pine martens are not confined to the fringes of the UK as was assumed, but that they have been living a secret life under our noses for decades.

http://www.guardian.co.uk/environment/2010/jun/04/pine-martens-uk-comeback

---

Slugs

At the end of May 2000 Mrs Ravenscroft of Little Comberton found three very large slugs in a bowl. These were passed to P. F. Whitehead for examination. He identified them as the limacid slugs, *Limax maculates*. These are rarely recorded in Britain. Two fell prey to a blackbird that probably had no idea of their significance. On 2 September 2001 a second site containing limacid slugs was found in Pershore.[84]

Snails

On 26 August, 1996, a distinctive snail marked with a pale line around the outer rim of the shell and a conical shape above was found in a garden at Little Comberton near Pershore. This was a specimen of *Hygromia cinctella*. It is normally found in South East France and this was the most northerly sighting.[85]

Snakes

In 2004 Gloucestershire Police issued a warning after a sighting of a six foot long snake in the village of Slimbridge. The large grey snake was seen on 10 August in a field behind the church of St John the Evangelist. On 30 September 2009 the *Gloucestershire Echo* reported that a violin expert had found the rattle from a rattle snake inside a violin. In some counties it is customary for rattlesnake tails to be placed inside violins when they are made

Soudley Beast

On 1 March 2007 *The Forrester* published a letter from a reader whose name and address were withheld. It told of the correspondent's walk around Soudley Pond in May 2003 and an encounter with a fat brown animal that was flapping wings which seemed too short to hold it up. The creature disappeared suddenly. One suggestion from a reader the following week was that the animal was a Mandarin duck which had strayed from the nearby Dean Heritage centre. However the Gloucester Paranormal and Fortean Investigation (GPFI) team had received an email similar to the original letter just prior to its publication. This contained three differences. Firstly it identified the name of the witness. Secondly it described the creature as four foot tall and thirdly it states that the witness was unable to see the face. GPFI regarded the report as a hoax.

Spiders

In 1998 eleven new species of spider new to Worcestershire were recorded.[86] In November 2001 a funnel web spider, one of the world's most venomous, was found in a glass jar by a woman walking in a forest near Kidderminster. The spider died of dehydration.[87]

On 19 May 2009 the *Gloucestershire Echo* reported that the false widow, a cousin of the deadly Black widow spider was establishing colonies in Gloucestershire. David Haigh, a spider recorder, said that there were two recorded instances in the county. One was in a shed in Tewkesbury in June 2007 and the other in July 2008 in Longney. A woman in Cumbria had been bitten in her home by a false widow. On September 6 2008 *The Gloucestershire Echo* reported that a colony of *Segestria florentia* had moved to Gloucestershire. This is the biggest European segestrid spider, reaching a body length of twenty two millimetres for males and fifteen for females. Forty were discovered living in a wall in a house in Ducie Street, Tredworth. It was not known how they had managed to spread so far inland.

Terrapins

In March 1999 the Worcestershire reptiles and amphibians group visited some sites in Redditch where they saw a terrapin. A few days later Mike Sutton, from the group, identified them as the common red-eared terrapin. These were imported to Britain as pets from the United States and Malaysia but such imports are now banned. Mr Sutton reported that the group were aware of seventeen terrapins in Worcestershire.[88] John Day reported seeing a red-eared terrapin in Norton in April 1993. When he returned to the same site in August 1999 the same, or perhaps another, terrapin was in the same position.[89] In 2002 the Gloucestershire Naturalists Society received a report of dark-coloured terrapins in Adlestrop Lake. There were also sightings in the Sharpness Canal (see Appendix One) which may or may not be relevant to the above mentioned sighting of a crocodile in that canal.[90]

Tortoises

In September 2006 Gloucestershire Police recovered two very young Mediterranean Spur-Thighed tortoises, *Testudo gracea sp*, that had been taken from the wild and sold illegally in a Cirencester pub. The buyers, who were unaware of the laws controlling the sale of protected species, cooperated with the Police investigation. The tortoises were given a new home by the RSPCA.[91]

Wallaby

On Thursday 9 May 1996 a sixteen year-old boy was returning home to the Hucclecote area of Gloucester when he saw an animal resembling a kangaroo bounce in front of him. The following night a similar animal was seen in the grounds of the Nuclear Electric head office in Barnwood. The *Gloucester Citizen* set up a telephone line, Wallaby Watch. One man who called this line was Dean Woodcock who was walking his dog in Buckholt Wood around 11.30 on 13 May 1996 when he saw the animal. There were two further sightings, one by a motorist in Barnwood on the evening of 13 May 1996 and another on the following day by a driver passing the Cross Hands public house at Brockworth. [92] A colony of red-necked wallabies, *Macropus rufogriseus*, lived in the Peak District after the second world war. They had belonged to Captain Courtney Brocklehurst, who was killed during the war. In 2000 they were thought to have become extinct but there have been sightings since and also reports from other counties.

To close this section of oddities I refer to a report in the *Gloucestershire Citizen* on 13 December 2008 that an 8lb 7oz bunny had been found in a garden in Knowles road, Tredworth. Snowy was so big that the RSPCA struggled to fit him in a cat carrier. His owner was never found but, like the wild boar, porpoise, pine marten, coypu and capybara he was definitely real.

SIX

Baffling Birds

O rnithologists are able to view a rich variety of bird life in Gloucestershire and Worcestershire, with many nature reserves that capitalise on the scenic locations. The international centre for birds of prey is located near Newent and a falconry centre is at Hagley. Many wild birds are attracted to the Slimbridge Wildlife Centre now owned by the Wildfowl and Wetlands Trust. Sir Peter Scott, whose father earned posthumous fame for his bravery in the ice, founded the centre in 1946. The reserve for waterbirds covers 120 acres. Although there are eight other wetland centres around the country. Slimbridge is the only place in the world where all six species of flamingos can be seen. On 29 April 2010 local and national newspapers reported the birth of eight baby cranes at the centre. Cranes died out in England during the 1600s but bred in Norfolk in 1982, and in most subsequent years.

The plan is for humans to teach these youngsters how to survive in the wild and then to release them in Somerset where it is hoped a viable population could develop.

Some of the other unusual, rare and endangered birds seen at Slimbridge, or elsewhere in Gloucestershire and Worcestershire, are documented below. The Rare Bird Alert website contains a full searchable national database and there are several birding websites, for each county, which contain more detail plus information on other species not mentioned here. These include the Birds of South Gloucestershire and Worcestershire Birding. Unless otherwise stated the data given below comes from Rare Bird Alert or the two mentioned websites.

Alpine Swift (*Tachymarptis melba*)
The Alpine Swift is a small bird, similar to a large House Martin, but from a different species. They breed in mountains from Southern Europe across to the Himalayas. There were four sightings in Worcestershire between 1973 and 1997 with the most recent being at Cookley on

28 April. In Gloucestershire there are two records, from 1970 and 1977. The second was at South Cerney Sewage farm on 1 June.

American Wigeon (*Anas americana*)
This is common in North America and Canada but rare in Europe. One was seen at Westwood Pool in Worcestershire on 22 January 2008.

Aquatic Warbler (*Acrocephalus paludicola)*
This is a rare migrant to the United Kingdom, with approximately forty being seen each year. A specimen was trapped, ringed and released at Littleton-on-Severn in 1976. In Worcestershire one was seen at Oakley Pool on 19 August 1983. Another appeared at Slimbridge at 15:25 on 10 September 2004.

Arctic Redpoll (*Carduelis hornemanni*)
The Artic Redpoll is a species of finch that breeds in tundra. Approximately two visit Britain every year. Four were seen in Wyre Forest between 14 and 26 January 1996 and another was present in Habberley Valley between 30 December 2001 and 24 January 2002. There has just been one sighting in Gloucestershire, at Highnam Wood between 16 and 21 February and 20 to 21 March 1996.

Avocets (*Recurvirostra avosetta*)
Avocets are waders, typically found in warm climates. They became extinct as a breeding species in Britain in 1842 but started breeding on reclaimed land near the Wash, a large estuary in East Anglia, during the 1940s. The Wash had been returned to salt marsh to try and deter German invaders. In 2003 Avocets began breeding for the first recorded time in Worcestershire at Upton Warren.[93] They are the second most common breeding wading bird at Slimbridge, after Lapwing.

Baltic Gull (*Larus fuscus*)
Nine Baltic gulls have been reported in South Gloucestershire. Five were reported from Littleton on Severn on 31 March, 1978. They were followed by two in Rangeworthy in 1979 and two near Oldbury on Severn in April 1981.

Baltimore Oriole (*Icterus galbula*)
This is the state bird of Maryland and has given its name to a baseball team, the Baltimore Orioles. The first sighting in Gloucestershire was in Horsley on 4 May, 2001. The Gloucestershire Naturalists Society noted that this was one of five new birds found in the county that year. It is very rarely seen in Britain.

Bewick's Swan (*Cygnus columbianus*)
The Bewick's Swan is the smallest of Britain's three swans. They breed on the Russian tundras and only visit Britain between October and March. Slimbridge is one of the places that attracts them. Sir Peter Scott began a study of the swan in 1964 when some of the wild swans arrived at a lake in the Rushy Pen at Slimbridge. As each individual swan's bill pattern is dif-

ferent it is possible to document the movements of individuals. The study continues today, in respect of the three hundred or so that arrive at Slimbridge each year. Around half of them were previous visitors. Bewick's Swan also visits Worcestershire on a regular basis. A review of 2007, published on the Birding Worcestershire site, noted twelve sightings in the county for that year.

Black-throated Thrush (*Turdus atrogularis*)
This bird is a migratory Asian species, and is regarded as rare in Britain. The first winter male in Worcestershire was found in a suburban garden in Webheath, in January 1996. According to the website of bird photographer, Jonathan Wasse, it attracted several bird-spotters who annoyed local residents.

Black Winged Stilt (*Himantopus himantopus*)
This widely distributed wading species is found across Africa and central Asia and is a rare visitor to the UK. Two sightings are noted on the Worcestershire Birding website, at Larford between 14 and 16 June 1985 and at Upton Warren between 21 and 22 May 2006. Upton Warren is a large nature reserve, managed by the Worcestershire Wildlife Trust. It is situated between Droitwich and Bromsgrove, divided by the river Salwarpe.

Cattle Egret (*Bubulcus ibis*)
The Cattle Egret evolved on the African savannah and, along with the Arctic tern, is the only bird to have expanded its range to reach all seven continents. One was present intermittently in Besford, near Pershore between 25 October and 26 December 1993. The following day it was seen at Bredon's Hardwick. There have been six sightings in Gloucestershire between 1974 and 2008, with the latest being in Frampton-on-Severn between 17 April and 22 June. In that year the species successfully bred for the first time in Britain, in Somerset.

Collared Pratincole (*Glareola pratincola*)
This little wader is found in the warmer parts of Europe, , southwest Asia and Africa. The annual report of the West Midlands Bird Club noted that one was seen at Bredon's Hardwick on 4 May 1994, the first sighting in the region. It flew into Gloucestershire.

Cory's Shearwater (*Calonectris diomedea*)
As this species usually breeds in the Mediterranean, it is unusual to see it in the UK. On 26 November 2000, just after 09:00, Brian Lancaster, Paul Bowerman and Dick Reader saw a large Cory's Shearwater moving up the channel at Severn Beach, accompanied by a couple of herring gulls. Later that day it was seen in several locations in North Somerset.[94] Although this bird is often seen in Cornwall, in July and August, it rarely ventures further inland.

Dartford Warbler (*Sylvia undata*)
This is basically a Mediterranean species, and southern England is on the utter edge of its range. Five Dartford Warbler's have been reported in South Gloucestershire, the first being on 5 October 2005. A single wintering bird was found on the Lickey Hills, Worcestershire in 1995 and another at Castlemorton Common two years later.

Egyptian Goose (*Alopochen aegyptiacus*)
The British population of this species dates back to the 18th century, though it was only formally added to the British list in 1971.Three have been reported in South Gloucestershire, the first near Oldbury Power station on 12 February 2002. The second was seen on the south shore of Severn Beach, near Oldbury yatch club on 4 March 2006. The third was photographed at Heneage Court Pools near Falfield on 23 April 2006. In Worcestershire five Egyptian Geese appeared at Upton Warren on 18 February 2007. One also arrived at Lower Moor on 25 February 2006.

Leucism is a rare condition in animals and birds where the pigmentation cells fail to develop properly. This can result in white patches appearing or, occasionally, the creature being completely white. In June 2008, Ray Sherman, photographed a leucistic Egyptian Goose at Slimbridge.[95]

European Bee-eater (*Merops apiaster*)
This species breeds in southern Europe and in parts of north Africa and western Asia. It is strongly migratory, wintering in tropical Africa, India and Sri Lanka. This species occurs as a spring overshoot north of its range, with occasional breeding in northwest Europe. It is a scarce visitor to the UK, although it was first recorded in 1668. The only record of this bird in South Gloucestershire was on 13 May, 2007 when it flew over Severn Beach, calling out. Bee-eaters devour wasps as well as bees. There have been three appearances in Worcestershire, at Kidderminster on 2 September 1955, Redditch between 22 and 27 September 1970 and Upton Warren on 28 May 2007.

European Roller (*Coracias garrulus*)
The European Roller, *Coracias garrulus*, is the only member of the roller family of birds to breed in Europe. Its overall range extends into the Middle East and Central Asia and Morocco. The first Roller in Gloucestershire was seen at Andoversford on 20 July 2001.

Gadwall (*Anas strepera*)
The Gadwall is a scarce-breeding bird and winter visitor in the UK, and was a rare bird in Gloucestershire but is now found at Slimbridge. In Worcestershire a few pairs breed at Grimley Old Working, which is just off the A443.

Glaucous Gull (*Larus hyperboreus*)
Glaucous gulls are the largest gulls in Europe. They breed in the Arctic. Just seven sightings have been recorded in South Gloucestershire between 1840 and 1991 and one of these didn't land. In January 2009 one landed at Slimbridge.

A spokesman credited the centre's staff who had made the enclose look like authentic tundra.[96]

Great Bustard (*Otis tarda*)
The Great Bustard was hunted to extinction in the mid nineteenth century with the last

Worcestershire record being in circa 1825 and the last in Gloucestershire in 1891 until one was seen flying south over Leckhampton and Leckhampton Hill on 24 May 1977. The male is one of the heaviest birds in the world. A reintroduction project in Wiltshire, which has two full-time employees, aims to create a stable population.

Great Grey Shrike (*Lanius excubitor*)

These are the largest of the European shrikes. Their Latin name means sentinel butcher, referring to their habits of storing food animals by impaling them on thorns and using exposed high areas to scan for prey. Three have been reported in South Gloucestershire, the first being in January 1891 at Dyrham Park and the most recent at Filton Golf Course on 6 April 1983.

Castlemorton Common in Malvern is regarded as the best place in the midlands to see a Great Grey Shrike. They were regular visitors during the 1970s then were not seen until 1989 and disappeared again after 2000.[97]

Great Northern Diver (*Gavia immer*)

These normally arrive on Britain's coasts during the winter. Two appeared near Fairford, on pit 125 at the eastern end of the Cotswold Water Park on 12 December 2009. The first sighting in South Gloucestershire was on 14 December 1983 and nine others have been seen in the area since. There have been fourteen sightings in Worcestershire, with the most recent being at Bittell Reservoir on 7 November 2001.

Great Skua (*Stercorarius skua*)

This large seabird breeds in the North Atlantic, including northern Scotland, The Faroes, and Iceland. There have been four sightings in Worcestershire between 1971 and 2000. It is a rare migrant through Slimbridge.

Great White Egret (*Ardea alba*)

This magnificent bird is found in most of the tropical and warm temperate regions of the globe. In recent years it has been a vagrant in the UK. The first sighting in Worcestershire was at Westwood Pool in September 1999.

One was also observed at Ashleworth Ham reserve between 17 and 19 March 2010. Several birding websites for Gloucestershire published photographs. This bird was colour ringed and its movements traced back to Brittany and then Lancashire, where it spent the winter.

Green Winged Teal (*Anas carolinensis*)

There have been three sightings of this North American species in Worcestershire at Bredon's Hardwick between 26 December 1992 and 5 January 1993, Grimley on 18 January 2003 and Bredon's Hardwick on 24 April 2004. The 2001 Ornithological report for the Gloucestershire Naturalists Society noted that one had wintered at Slimbridge in that year.

Honey Buzzard (*Pernis apivorus*)

This bird is a passage migrant in most of Europe, but winters in Africa. Two were found at

Totworth in 1879. There were unconfirmed rumours of a honey buzzard in the vicinity of Bredon School in 2009. Sixteen sightings have been recorded in Worcestershire, most recently at Upton Warren between 28 and 29 May 2004.

Isabelline Shrike (*Lanius isabellinus*)
This bird lives in south Siberia and Central Asia, and is a very rare vagrant to these shores. A first winter Isabelline Shrike for Gloucestershire was seen in the Cotswold Water Park, pit 79 on 28 October 2001.[98]

Lapland Bunting (*Calcarius lapponicus*)
This is a winter visitor, and scarce visitor. They seldom linger here during the winter and only very rarely drop in during their return migration in spring. Until 1892, when considerable flocks invaded Norfolk and Suffolk, they had scarcely ever been seen in East Anglia or elsewhere in Britain; but from that year onwards their migratory habits changed and they are now seen almost every autumn. This species was first reported in South Gloucestershire in 1981 and there have been nine sightings since. The first record in Worcestershire was on 7 and 8 October 2007 at Upton Bittell Reservoir.[99]

Lesser Scaup (*Aythya affinis*)
This duck breeds in north and central North America, wintering in northern South America. As of 10 November 2008 there were 127 accepted records of the Lesser Scaup in Britain. Three of these sightings were in Gloucestershire, one in 1994 and two in 2008, at Lydney lakes between 31 October and 17 November and one between 22 November and 7 December at Frampton-on-Severn. This was thought to be the same individual. One was seen at Grimley and Westwood Pool in February and March 2006.

Lesser Spotted Woodpecker (*Picoides minor*)
This bird was seen at Little Stoke on 1 February, 1942 and has appeared more frequently since. It is a priority species in the UK biodiversity action plan due to populations being threatened in both UK and overseas.

Little Auk (*Alle alle*)
This Arctic bird is a passage winter visitor, and not the sort of bird one would expect to find in Gloucestershire or Worcestershire. On the morning of 3 November 2003, following high winds, a little Auk was retrieved from the *Mabey and Johnston Ltd* factory in Lydney and taken into care. Twenty five Little Auks have been reported in South Gloucestershire, between 1841 and 2009 with seven being in the last year.

Little Bunting (*Emberiza pusilla*)
This bird which is normally found in northern Eurasia is a scarce visitor to the UK. There have been two sightings in Worcestershire. The first was at Eckington between 25 and 26 May 1994 and the second, involving three birds, was at Caunsall between 5 February and March 2005.

Long-billed Dowitcher (*Limnodromus scolopaceus*)

These waders breed in the high tundra of Siberia and the far north of North America, but winter as far south as Central America. They are rare visitors to Europe. In Worcestershire this was seen at Westwood Pool between 9 and 20 October 1990 and at Bittell Reservoir between 25 September and 3 October 2006. There have been two sightings in Gloucestershire in 1984 and 1985, with the second being a juvenile at Slimbridge.

Long-eared-owl (*Asio otus)*

Long-eared owls are a British resident, but have a reputation for being difficult to see. They weigh less than 300 grams. In the past they have bred near Oldbury power station and Warmley Forest park. Around forty were reported in South Gloucestershire between 1897 and 2009.

Long-tailed Skua (*Stercorarius longicaudus*)

Another Arctic bird which winters in the South Pacific and SE Atlantic. It has occasionally bred in the UK. The smallest of the Skuas was first seen in South Gloucestershire in 1981. Five sightings have been reported since, all around New Passage. In Worcestershire it was recorded at Upton Warren on 8 August 1987.

Nightjar (*Caprimulgus europaeus*)

The Gloster Nightjar was a RAF fighter aircraft of the early 1920s and provides a suitably bizarre lexilink with which to begin this section.

In Gloucestershire nightjars which were close to extinction are enjoying a renaissance. They are nocturnal, or crepuscular, birds with long wings, short legs and short bills. The erroneous belief that they suck milk from goats has led to their nickname of "goatsuckers." Their churring call sounds like a lawnmower. In 1982 there were twenty churring males in the Forest of Dean, reduced to twelve in 1992. However the opening up of glades that they need for nesting and feeding saw numbers increase.[100]

Osprey (*Pandion haliaetus*)

Until the 1950s Ospreys were rare visitors to Britain and did not nest here. Now they nest in Scotland, and occasionally in England and Wales, but can also be seen flying through. On 9 April 2000 at 16.30 Nick and Georgie Jones saw an Opsrey in a tree near Whitecliff Farm in Coleford.[101] There was also a sighting near the Severn Beach that year. In total 21 have been reported in South Gloucestershire, between 1860 when one was killed whilst fishing at Totworth Court Lake and 2009. Apart from the fishing victim none stopped.

Pallas's Warbler (*Phylloscopus proregulus*)

This is a leaf warbler which breeds in southern Siberia (from Novosibirsk Oblast east to Magadan Oblast), northern Mongolia, and northeastern China. It is strongly migratory and winters mainly in subtropical southern China and northeastern Indochina, but also in small numbers in western Europe. They are even smaller than a goldcrest and weigh 3-7g. By the time they arrive in Britain they will have travelled over 3,000 miles. One was seen at Westwood Pool on 17 November 1987.

Peacock (*Pavo cristatus*)

On 6 May 2004 the BBC reported that fourteen peacocks were making a nuisance of themselves in Cookley, Worcestershire by destroying roofs and gardens and generally making a mess. Nobody had come forward to claim ownership of the birds. The website of South Gloucestershire Council contains an information sheet for peacock owners. This was prompted by complaints about the birds being noisy and trespassing.

Peregrine Falcon (*Falco peregrinus*)

On 31 August 2009 the *Gloucestershire Echo* reported that growing numbers of birds of prey were threatening the sport of pigeon racing. During the second world war the Ministry of Defence ordered the destruction of peregrine falcons because pigeons were essential to the war effort. The falcons survived and two nested in St Andrew's Church, Worcester in 2007 and 2008.

Pied Wheatear (*Oenanthe pleschanka*)

A male was seen near Bredon Hill on 5 November 2005. This species is usually found between South Eastern Europe and China, rarely venturing into Western Europe. It had never been seen previously in a inland British county. [102]

Pigeon (*Columba livia* f. *domestica*)

A large fledgling, known as Big Boy, puzzled staff at the *Oak and Furrows* rescue centre in Somerford Keynes because of his enormous size. Fortunately his owner read a newspaper article and the family reunion was reported by the *Stroud News and Journal* on 11 June 2009. Big boy and his father, also unusually large, came from a rare breed known as a runt.

Pine Bunting (*Emberiza leucocephalos*)

An Asian species that is a rare vagrant to western Europe. There have been two sightings in Worcestershire, at Bibbey's Hollow between 6 and 20 February 1996 between 15 and 24 January 2004 at New Farm.

Puffin (*Fratercula arctica*)

A puffin was found at Filton Airfield in 1935. Others have been seen around Severn Beach in 1951, 1997 and twice in 2002. In Worcestershire one was seen at Hagley in July 1936 and another at Redditch on 30 August 1963.

Purple Heron (*Ardea purpurea*)

This is a species normally found in southern Europe, southern and eastern Asia and parts of Africa. It is considered a passage visitor to the UK, but bred here for the first time in 2010. There have been two sightings in Worcestershire. The first was at Upton Warren on 5 May 1982 and the second at Upton Warren and Oakley Pool between 29 June and 1 July 2000.

Purple Sandpiper (*Erolia maritime*)

This Arctic species is a winter visitor on the east and south coasts. It has bred in small numbers in Scotland. Four were seen at Bittell Reservoir between 31 October 1940 and 24 August

1988.

Red Footed Falcon (*Falco vespertinus*)
This eastern European bird is a rare vagrant to the UK. In Worcestershire there was a sighting at Westwood Pool on 15 May 2001. In Gloucestershire there were six sightings between 1972 and 1994, with the last relating to the Cotswold Water Park.

Red Kite (*Milvus milvus*)
The red kite was on the verge of extinction at the start of the twentieth century but recovered to be voted bird of the century in 1999 although it remains rare. One was seen flying near Tockington on 3 March, 1957.

Red-rumped Swallow (*Cecropis daurica*)
This widely occurring species is found from southern Europe to Japan. The first record in Gloucestershire was from Slimbridge on 5 April 2001. The second, and last to date, relates to two specimens seen in the Forest of Dean in 2004. In Worcestershire there were sightings at Upton Warren between 1 and 2 May 1992 and Westwood Pool on 17 April 2001.

Rhea (*Rhea spp*)
The *Worcester Evening News* reported on 7 February 2009 that a six foot Rhea had been running loose in Pershore three days earlier. It was first seen in Station Road before being captured in Hurst Road. The bird, named Charlie, had become frightened at a poultry farm in Drakes Broughton and escaped.

Richard's Pipit (*Anthus richardi*)
This Asian species is a rare vagrant in Britain. According to the 2001 Ornithology report from the *Gloucestershire Naturalists Society* a Richard's Pipit was discovered lurking in fields between Frampton and Cambridge. It remained there into the following year. The first record in South Gloucestershire was on 27 November 1993 when a winter adult was found at Aust Warth and remained there for three days. Four others have been spotted since, in 1996, 2006, 2007 and 2009.

In Worcestershire there were sightings at Upton Warren on 7 October 1967 and Grimley on 7 October 2007.

Ring-necked Duck (*Aythya collaris*)
This is a North American species, which is only a scarce visitor in the UK. On 3 February 2010 the BBC's website reported that the ring-necked duck had been seen at Slimbridge for only the fourth time in 50 years. The first sighting of the duck in Europe was made by the wife of Peter Scott in 1955. That sighting was the first conclusive proof that these birds did journey across the Atlantic.

Four have been reported in South Gloucestershire, two at New Passage in 1998 then two separate sightings in 2002 at Oldbury power station in April and Orchard Pools near Severn Beach in November. There have been three sightings in Worcestershire, with the first being at Bittell Reservoir on 28 March 1990.

Rose Coloured Starling (*Sturnus roseus*)
This is normally a bird of eastern Europe and southern Asia. It is a scarce visitor to the UK. There have been four sightings in Worcestershire. One was shot in Powick in August 1855. A second appeared in Aston Somerville between 23 and 25 August 2002. The third was in Tenbury Wells between 6 and June 2003 and the fourth was at Callow End in April 2005.

Ruddy Duck (*Oxyura jamaicensis*)
Sir Peter Scott brought three pairs of the North American Ruddy Duck, *Oxyura jamaicensis*, there in the 1950s. The population grew to several thousand in Britain during the 1990s and reached twenty European countries. In Spain hybrids of the ruddy duck and their cousin, the white headed duck *Oxyura leucocephala* began to emerge. To protect the white headed duck DEFRA began culling the ruddy duck in Britain. Environmental groups criticised this, not only on moral grounds, but also because DEFRA's figures showed that £4.6 million of taxpayer's money had been spent killing a total of 6,200 birds.[103]

Rustic Bunting (*Emberiza rustica*)
It breeds across northern Europe and Asia. It is migratory, wintering in south east Asia, Japan, and eastern China. It is a rare wanderer to western Europe. One was seen at Upton Warren on 7 November 1987.

Sabine's Gull (*Xema [or Larus] sabini*)
This Arctic bird is a passage visitor to the UK. A post on the *Gloucestershire Naturalist's website* on 6 November 2003 stated that a juvenile was seen near Slimbridge earlier that week, following some extreme winds in the area.

Savi's Warbler (*Locustella luscinioides*)
This is a rare visitor from western Asia. A male was seen at Frampton-on-Severn from 20 May to 1 June, 2001.

This is the only record in Gloucestershire. A sound recording was taken. The *Worcestershire Birding* website notes four sightings in the county between 1985 and 1999. However the Rare Birds Alert only confirms the last of these where a male singing was seen at Oakley Pool in Droitwich.

Serin (*Serinus serinus*)
This bird - the smallest European finch - is a scarce visitor from southern Europe and north Africa. Two of the species were seen in Evesham on 17 and 18 June 1978 and there were two sightings in Wilden on 31 January 1982 and 1 April 1982.

Shorelark (*Eremophila alpestris*)
The Shore Lark breeds across much of North America from the high Arctic south to the Isth-

mus of Tehuantepec, northernmost Europe and Asia and in the mountains of southeast Europe. A shorelark was seen on 31 January 1960 at Severn Beach and stayed until 4 March. There was a further sighting, also at Severn Beach in 1977 and six others in South Gloucestershire in between January 2004 and January 2010.

Squacco Heron (*Ardeola ralloides*)
One visited a small fishing pool at Broadoak Trout Lakes near Hanley Swan on two occasions in June 2007. This was a first for Worcestershire as the bird rarely ventures north of its breeding range in Southern Europe and the Middle East.[104] There were just two sightings in Gloucestershire in 1867 and 1997. The most recent was in the Cotswold Water Park between 24 June and 1 July.

Stone Curlew (*Burhinus oedicnemus*)
This bird bred in Worcestershire until circa 1840. There were four sightings between May 1991 and May 2008. It is a rare migrant breeder in the UK.

Storm Petrel (*Hydrobates pelagicus*)
It breeds on inaccessible islands in the north Atlantic and western Mediterranean, with the core population in western Ireland, northwest Scotland and the Faroe Islands, where the worldwide biggest colony breeds on the island of Nólsoy. Four have been reported in Worcestershire since November 1936 but three were found dead. The exception was at Bewdley on 13 July 1968.

Whiskered Tern (*Chlidonias hybridus*)
This rare vagrant is usually found in the warmer parts of Europe and Asia. There was a sighting at Bredon's Hardwick on 1 May 1994. There are three recorded sightings in Gloucestershire, with the most recent being at Frampton-on-Severn in 2008.

White rumped Sandpiper (*Calidris fuscicollis*)
These are rare vagrants to the UK. Their breeding habitat is the northern tundra on Arctic islands in Canada and Alaska. They nest on the ground, usually well-concealed in vegetation.Six have been reported in Gloucestershire, with four in South Gloucestershire where the first sighting was on 12 September 1985 at Severn Beach. It stayed overnight. Others were seen at Severn Beach in 1995 and 2001, with another at Aust Warth in 1998.

In Worcestershire a juvenile was seen at Bredon's Hardwick between 15 and 16 September 1996.

White Spotted Bluethroat (*Luscinia svecica cyanecula*)
This bird is found in central and eastern Europe across Asia, and with a small population in Siberia. It is a passage visitor to the UK and occasionally breeds here. A male white-spotted bluethroat was spotted at Slimbridge on 23 April 2010.

White Stork (*Ciconia ciconia*)
This iconic bird usually breeds in northern and central Europe, and Asia Minor, wintering in

India and north Africa. It is a scarce visitor to the UK, but was once resident. The last attempted breeding was in 1416. One was reported on 23 May 1971 from the Hallen/Compton Greenfield area, 1971.

White tailed Eagle *(Haliaeetus albicilla)*

This eagle, also known as the sea eagle, was once declared extinct in Britain, although more recently reintroductions have taken place in Scotland. According to the *Western Daily Press* of 13 April 2008, Peter Alder, an 87 year-old former warden at Slimbridge, saw the eagle on the previous day whilst with a friend, Nick Gardiner. The first record in South Gloucestershire was at Dodington Park in December, 1871.

Whooper Swan *(Cygnus Cygnus)*

Only occasionally breeding in the UK, this species has in recent years become a more common winter migrant to the eastern part of our country. Three adults roosted on floodwater at Hasfield Ham and then flew off to land in a field halfway between Wainlodes and the A38 on 3 March 2010. In total fourteen were reported in South Gloucestershire between 1983 and 2009.

Wilson's Phalarope *(Phalaropus tricolour)*

This bird was seen at Slimbridge feeding with ducks on 8 November 2009. It is the largest of the phalaropes and breeds in North America, rarely venturing to Western Europe. There had only been three previous sightings in Gloucestershire between 1976 and 1990. In Worcestershire it was recorded at Upton Warren on 14 September 1985 and, in the same location, between 23 and 26 September 2007.

Woodlark *(Lullula arborea)*

The woodlark is now rare in Gloucestershire. Although there have been isolated sightings, such as near Crabtree Heath in 1998, there are no known breeding populations. Accordingly the Gloucestershire Biodiversity Action Plan set a target to restore a breeding population by 2010.

Wryneck *(Jynx torquilla)*

In 1829 the Wryneck was considered to be in decline in Gloucestershire but large numbers visited in 1902, mostly in the Severn Vale and Forest of Dean. There were no records of it breeding.[105] In Worcestershire the population declined between 1912 and 1934, reducing it to the status of passing visitor.[106]

Whilst researching this chapter I became aware of many people who spend an extraordinary amount of time watching birds and recording the details. They may not see themselves as Cryptozoologists but have added much to our knowledge of the birds that live around us and deserve credit for their dedication.

Picture of a Great Bustard, from a 1905 German Encyclopaedia, downloaded from http://commons.wikimedia.org/wiki/File:Greatbustard.jpg, 31 May 2010.

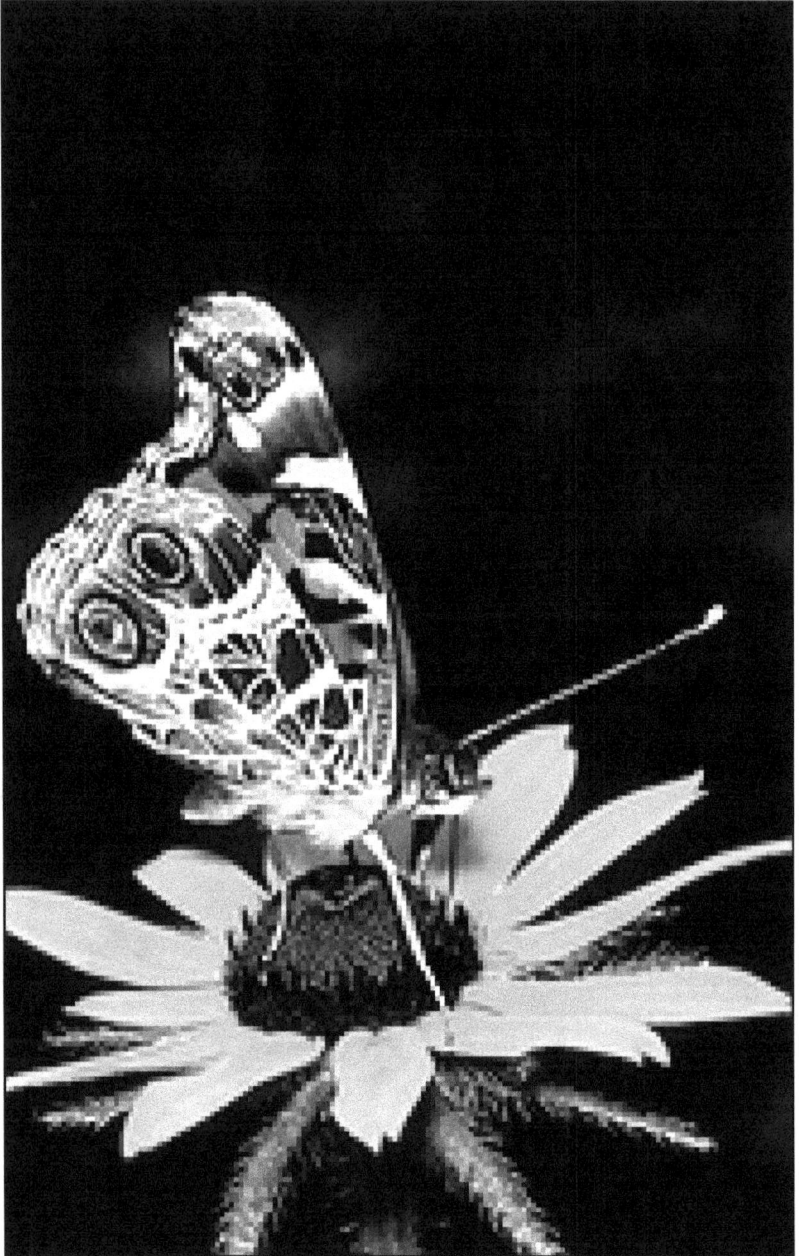

Picture of a Painted Lady Butterfly, once described by a Victorian naturalist in Gloucestershire, image in the public domain downloaded from http://www.junglewalk.com/photos/Butterfly-pictures-I2842.htm, 31 May 2010.

SEVEN

Incongruous Insects

The vast majority of cryptozoological discoveries are in the insect world but these rarely receive the publicity afforded to less credible sightings of large mammals. This is probably because the yeti, big cats and bigfoot are topics that can be relied upon to attract publicity and sell books. Millipedes and ants do not attract the same general interest and Entomology is often, regrettably, a closed science. Some of the insect discoveries in Gloucestershire and Worcestershire are summarised below in alphabetical order by species.

Ants

On 1 August 2009 the *Daily Telegraph* reported that colonies of *Lasius neglectus*, the so-called Asian super ant had been found at Hidcote Manor near Chipping Campden in Gloucestershire. An estimated thirty five thousand of the ants were discovered in an electricity junction box after workers found dead specimens around cables. This was thought to be the first recorded sighting in the United Kingdom, although they have been spotted in mainland Europe. The fire ants are naturally drawn to electrical currents and swarms can cause blackouts. They look the same as the common black garden ant but have a different social structure and patterns of behaviour.

Bees and Wasps

Bee-wolves (genus *Philanthus*), are large bright yellow insects which attack honey-bees before burying their paralysed bodies. In August 1997 Brett Westwood and Harry Green found bee-wolves in soft sand near the Stourport Road at Hartlebury Common. Until the mid 1980s they were restricted to a single colony on the Isle of Wight. Gradually they began to spread across the mainland.[107] The population has now expanded to sandy soils at Norchard Farm but specimens have also been seen on normal agricultural land.

On 14 August 2001 Kevin Mc Gee photographed a very dark wasp at Mill Meadow, Drakes

Broughton. George R Else at the Natural History Museum said that it was the darkest specimen he had seen of the male *Dolichovespula media*. The species was a new arrival to the British Isles. They usually hang their nests low on branches, making them more prone to disturbance from gardeners. Mr Mc Gee subsequently found a queen of the species at Tiddesley Wood on 20 October 2001. [108]

On 8 August 2008 the *Worcester Evening News* reported that Paul and Mike Southall had attracted one of Britain's most endangered species of bumble bee to Norchard farm near Stourport. The presence of *Bombus ruderatus* was confirmed by experts in a special conservation area created by the Southalls with the support of Sainsburys. In 1999 Geoff Trevis had reported seeing a specimen in Monk Wood. [109]

Beetles
In October 1997 David M. Green visited the Crews Hill Wood reserve near Malvern and collected some fungus. Three days later two shiny dark brown beetles were found in the container. These were identified as *Lycoperdina bovistae*. The species had never previously been reported in Worcestershire. [110]

In 2000 the first British record of *Ixapion variegatum*, a small weevil of mistletoe was found at the Brockhampton Estate near Bromyard in Herefordshire. In 2001 the same species was reported also at Pershore in Worcestershire and in South Gloucestershire. [111] On 28 August 2004 at Carpenters Farm, near Berrow, in Worcestershire the weevil was again recorded. In 2001 the rare bombardier beetle, *Brachinus crepitans*, was discovered near an old railway site in Honeybourne. [112] This beetle is between six and ten millimetres long, with a narrow thorax. When alarmed it ejects explosive pulsating jets of noxious gas from its rear end. The only previous record from Worcestershire was also from a railway track, at Brotheridge Green in the early 1970s.

On 26 June 2007 the *Worcester News* reported that a species of bug, *Eulecanium excrescens*, had arrived in Britain from Asia due to the increase in imported plants. It grows a grey sack around itself then lays thousands of eggs which led to the plant being quickly devoured. Other similar pests to migrate to Britain in recent years include the rosemary leaf beetle, *Chrysolina americana*, and the scarlet lily beetle, *Liloceris lilii.*

On 21 September 2007 the *Worcester Evening News* reported that beetle droppings found inside a tree at *Westons Cider* near Ledbury proved that the rare noble chafer beetle, *Gnorimus nobilis*, was alive. This insect has a metallic green body speckled with white. It has been rare in Britain for over a century and is usually associated with dead wood especially around old fruit orchards.

The *Stroud News and Journal* ran a feature on 28 January 2008 about a three year restoration project at a conservation centre in Haresfield, Gloucestershire, which involved thinning dense ash trees to prolong the lives of existing beech trees and allow new beeches to grow. The habitat is home to many rare insects, including the black-headed cardinal beetle, *Pyrochroa coccinea.*

In July 2008 a giant insect thought to be extinct in Britain was discovered on a pavement in Ashchurch just outside Tewkesbury. John Hawkins from Stourport handed the Capricorn Beetle, *Cerambyx cerdo*, to the Tewkesbury Council Environmental Health Department. It was then moved to the Stratford butterfly farm. Capricorn Beetles were believed to have vanished from Britain in the early eighteenth century and are rare elsewhere in Europe.[113] The insect was only identified through Mr Hawkin's persistence in capturing the specimen and reporting it to the right authorities. It was thought to have arrived in the country in imported oak timber. There are however suggestions that the species may be living in Britain. On 27 June 2006 the *Daily Mail* reported that a Capricorn beetle had been found at a furniture restorer's in Llanelli where it had crawled out of a pile of English Oak. On 28 May 2005 the *Daily Telegraph* reported that one had been found in a house in Warwickshire. It was also taken to the Stratford butterfly farm. The larvae of this beetle are considered a culinary delicacy in parts of the sub-continent.

Butterflies

The Essex skipper, *Thymelicus lineola*, was unknown in the Midlands until 1992 when it appeared in Warwickshire. It was first reported in Gloucestershire in 1996 and arrived in Worcestershire the following year.[114] On 7 August 2000 Chris Thompson discovered a Long Tailed Blue, *Lampides boeticus*, in his kitchen in Pershore. It rarely reaches Britain and this was the first sighting in Worcestershire.[115]

In August 2006 the Adonis Blue butterfly, *Lysandra bellargus*, reappeared in the Cotswolds, having been extinct for more than forty years. It recolonised National Trust sites at Rodborough and Minchinhampton Common.[116]

Painted lady butterflies, *Cynthia cardiu*, have been seen at the Slimbridge wildlife centre. They are not new arrivals to the county. John Leonard Knapp in his *Journal of a Naturalist* commented that visits by the painted lady were unpredictable. He informs us that in the summers of 1815-1818 very few moths or butterflies appeared in Gloucestershire but the painted lady became the most common butterfly seen in 1818. Some years earlier a quantity of earth had been raised to make a canal and the butterfly was found in abundance amongst the herbage that sprang from the new soil on the bank.[117]

On 24 November 2006 the *Worcester Evening News* reported that substantial populations of the Grayling butterfly, *Hipparchia semele*, had been found in Malvern. This rare species is grey and brown and usually has good camouflage.

On 19 August 2009 the *Guardian* reported that the Duke of Burgundy, *Hamearis lucina*, one of Britain's most endangered butterflies, had reappeared for a second generation for only the third time in over a century. Rodborough Common in Gloucestershire was the furthest north that a second generation had ever been recorded.

Caterpillars

On 8 August 2006 the *Worcester Evening News* reported that Ron Williams, of Vancouver Close in Worcester was bewildered by a two-inch long caterpillar found by his cat. The caterpillar was as fat as a small human finger. Various readers tried to identify it, with some saying that it was a lime hawk-moth caterpillar but no consensus was reached. Another possibility was an elephant hawk-moth caterpillar. On 20 May 2009 the *Gloucestershire Echo* reported that thousands of hungry caterpillars were munching their way across Churchdown. Cobwebs covered gardens in Vervain Close. The culprits were believed to be lackey moth caterpillars.

Centipedes and millipedes

On 9 September 2004 *The Sun* reported that Joy Jones from Charfield had been bitten by a centipede which she thought was the same one that she had killed and left for dead in her garden a year earlier. It had unwittingly been brought back from Santorini by her daughter, Bridget. The seven-inch long poisonous creature was taken to Bristol Zoo. On Friday 7 April 2009 David Scott-Langley discovered a pale yellowish millipede in a garden in Lower Oddington. The species was confirmed as *Cylindroiulus vulnerarius*, the first time it had been recorded in Gloucestershire.[118]

Crickets

April 2000 saw the first record in Worcestershire of the short-winged Conehead, *Conecephalus dorsalis* in Redditch. This is a small green cricket with a brown stripe on the back and brown wings. In September 2003 John Meiklejohn found a female long-winged Conehead, *Conocephalus discolour*, near Upton-on-Severn.[119]

Cockroaches

In September 1998 a lady in Defford opened a box of eggs purchased in Pershore and a large insect jumped out. This was believed to be an older nymph of the Australian Cockroach, *Periplaneta australasiae*. The eggs came from Newent but the origin of the insect was unknown.[120] In November 2007 health inspectors from Cotswold District Council shut down the Mogul Tandori in Cirencester after a customer found a cockroach in their takeaway.

Dragonflies

There are around thirty species of dragonfly in Great Britain. Two of them were new to Worcestershire when recorded there in 2000. They were the red veined Darter, *Sympetrum fonscolombei*, and the Lesser Emperor, *Anax parthenope*.[121] The lesser emperor usually thought of as a Mediterranean species was recorded in the Forest of Dean, Gloucestershire, in 1996. This was the first British record.

In June Ingrid Twissell reported that the rare Scarce Chaser, *Libellula fulva*, was discovered in good numbers by Dick Medley on the River Avon in Worcestershire. He contacted Mike Averill, the Worcestershire Dragonfly recorder who followed the Avon South and found the Scarce Chaser in Tewkesbury and Twyning.[122] This species is noted as scarce in the *British Red Data Book* on Insects.

Flies

Antomyia is a genus of fly in the family Anthomyiidae. The Natural History Museum has a single anthomyid specimen, the only one in Britain, identified as *Anthomyia bazini*. It came from Alfrick in Worcestershire in 1941. In 2002 David Gibbs found two new species of hoverfly for Gloucestershire. These were *Callicera aurata* and *Pipiza lugubris*.[123]

On 14 August 2003 the Tewkesbury AdMag published an article about a blackfly that fed on human blood being discovered on the outskirts of Bourton. Dr Adrian Pont, of the Oxford University Museum of Natural History, heard reports of the insects breeding in the Oxfordshire and Cotswolds areas in the March, so with a German colleague set up a study in the Oxford area coming almost as far as Bourton. This revealed large numbers of the blackfly. *Asilus Crabroniformis*, the hornet robber fly was found by Brett Westwood in Kidderminster in 1997 but has not been seen in Worcestershire since.[124]

Locusts

Locusts are probably best known in Britain as a Biblical plague inflected on the people of Egypt. On a personal level I recall my disgust at being asked to dissect one in a biology lesson. According to the *Leeds Mercury* of 15 October 1869 two Egyptian locusts were found in Birmingham. The following day the *Bristol Mercury* reported that a locust of immense size had been caught in Taunton. Charles Fort reported in his book, *Lo*, that on 8 and 9 October 1895 locusts appeared in large numbers in Pembrokeshire, Derbyshire, Gloucestershire and Cornwall.[125] Their origin is unknown.

Mosquitoes

On 24 August 2007 the *Daily Telegraph* reported that the Asian tiger mosquito was found in a garden in Cheltenham. According to the website Mosquito watch this was the first record of this mosquito species in Britain. The insect, which is about four millimetres long with tallow stripes, was recognised by Julian Berryman who had seen them elsewhere in Europe. This specimen was thought to have come across in a shipment by sea or air.

The Asian tiger mosquito can carry the West Nile virus which killed three hundred people in America in 2002. Given the recent panic about the relatively harmless swine-flu it is interesting that the presence of the mosquito did not draw further publicity. Perhaps this is because of doubts about the sighting. The Health Protection Agency, a subsidiary of the Department of Health, state on their website that there have been no confirmed sightings of this mosquito in the United Kingdom but also say that it is an insect which could potentially arrive here.[126] However a spokesman for the agency was quoted by the *Daily Telegraph* as saying that the two tiger mosquito sightings in Cheltenham were the first for the country.

Moths

On 3 June 2000 two males of the Pauper Pug *E. egenaria* were found in Shrawley Wood. This was the first sighting of the moth in Britain, outside of the Wye Valley and Norfolk.[127] In 2004 a new species was added to the Worcestershire list of micro-Lepidoptera and the status of another was changed from extinct to flourishing.[128] Field Maple which is the staple food of *Stingmella acrecis*, was found in hedges to the west of the Mythe Bridge. By 2005 it was confirmed that the moth was present in an area stretching from Durbridge Mill in the West to the Mythe in the east and from Staunton in the South to Gullers End in the North. Another micro-moth *Phyllocnistis saligna* was found in the leaves of a White Willow near Upper Lode in Tewkesbury. Prior to this the moth had been recorded at Stroud in 1953 but the last Worcestershire record was in 1873 and it was regarded as extinct. Between September 2004 and September 2005 Robert Homan amassed a total of seventy six monad records of *Phyllocnistis saligna* from the Three Counties. The records in Worcestershire range from Chaceley in the south to Salwarpe near Droitwich in the north and from Hollybush in the west to Evesham in the east.

Water Bugs

On 28 March 1997 Don Goddard discovered a bug previously unknown in Worcestershire in the River Teme at Ham Bridge between Martley and Clifton-on-Terme. *Aphelocheirus aestivalis* is the only representative of the family aphelocheiridae in Britain. It is about ten millimetres long and eight wide, dark olive green with pale legs.[129]

With this discovery we reach the end of our search for the mystery animals of Gloucestershire and Worcestershire. Some have been discovered, either by chance or as a result of diligent fieldwork by dedicated researchers. Others remain unidentified. If they do not exist then the mystery changes and becomes a question as to why someone has claimed that they do. There are no easy answers to that question.

Photo of some Fire Ants. A colony was found in Chipping Camden. Image in the public domain downloaded from http://www.junglewalk.com/photos/Ant-pictures-I9440.htm, 31 May 2010.

AFTERWORD

To conclude I note the saying that coincidences come in threes. In December 2008 my brother-in-law saw a big cat as he was driving on a remote road back to Castlemorton in the earlier hours of a wintry morning. After looking at various pictures, obtained during the course of my research for this book, he felt that the animal seen most resembled the jungle cat.

On 29 November 2009 I was driving along Green Lane, Bushley, with my wife when she told me to watch out for deer. I informed her that there were no deer in the area, as there was insufficient woodland. She replied that she had seen them on the road twice and on a third occasion when two came into our front garden and fled when the security light came on. Intrigued I discovered that there was a deer park near Eastnor Castle, about nine miles away. I am unsure if this is where the deer originated or if others have strayed into my village.

On the following day I was travelling by train to Birmingham and had just passed Bromsgrove station, when I saw a fox with a lamb in its mouth sitting in a field. If the train had been going at its usual speed my vision would have been restricted to a mere glimpse. Enough perhaps to believe that it was not a fox, but a cat. On other occasions I have seen black, domestic, dogs from a distance that might be mistaken for large cats. However I would not suggest misidentification as an explanation for all sightings of large cats. The sheer volume of reported sightings strongly suggests that there is something out there but only when a body is dragged in front of the media's cameras will the most famous mystery animal in the region be identified and accepted.

NOTES

1. BBC News, 11 January 2005, http://news.bbc.co.uk/1/hi/england/ gloucestershire/4166627.stm, accessed 29 October 2009.
2. List of Customs Seizures 2007, DEFRA, http://www.defra.gov.uk/wildlife-pets/ wildlife/trade-crime/documents/custom-seizure2007.pdfoaded 21 April 2009, accessed 3 November 2009.
3. *Hansard,* 21 July 2009, c1174.
4. 4. Cusdin, P. A., "The Keeping of Wolf-Hybrids in Great Britain", June 2000, Department for the Environment, Transport and the Regions, http://www.defra.gov.uk/ wildlife-pets/wildlife/protect/documents/dwa-wolfdogs.pdf, accessed 3 November 2009, p. 44 and 47.
5. Ibid. p. 47.
6. This link is no longer active.
7. *Hansard*, 6 November 2002, c374.
8. BBC News, 18 September 2006, http://news.bbc.co.uk/1/hi/uk/5353882.stm, accessed 2 November 2009.
9. Reported Sightings or signs of exotic species compiled by Defra Rural Development Service (2001 to September 2006) and Natural England (from October 2006) http:// docs.google.com/gview?a=v&q=cache:wi17rxoY55AJ:www.naturalengland.org.uk/ Images/exoticsstatssummary_tcm6-4154.pdf, accessed 5 November 2009.
10. Williams, Jan, "British Big-Cats-The Early Years," in Big Cats in Britain Yearbook 2006, edited Mark Fraser, p.12.
11. Matthews, Marcus., Big Cats Loose in Britain, (CFZ Press, 2007) p. 312.
12. Ibid., pp. 307-8.
13. BBC website, 03 May 2004, http://www.bbc.co.uk/gloucestershire/focus/2004/04/ bigcats.shtml, accessed 4 November 2009.
14. Greening, Mervyn, "Gloucestershire Naturalist's Society, Mammal Report, 1997", http://www.glosnats.org.uk/reports/mammal.php, accessed 10 November 2009.
15. *Hansard*, 2 February 1998, column 821.
16. British Big Cats Society, Press Release, 20 April 2004, http://www.britishbigcats.org/ news.php, accessed 3 November 2009.

17.   British Big Cats Society, http://www.britishbigcats.org/news.php, accessed 16 October 2009.

18.   House of Commons Debate, 20 March 1969, http://www.theyworkforyou.com/debates/?id=1969-03-20a.723.2&s=%22dangerous+wild+animals%22, accessed 3 November 2009.

19.   DEFRA,, http://web.archive.org/web/20061210055808/http://www.defra.gov.uk/wildlife-countryside/vertebrates/reports/exotic-cat-escapes.pdf, accessed 4 November 2009.

20.   *Hansard* 27 February 2003, c666w.

21.   DEFRA, http://www.defra.gov.uk/vla/vla/docs/vla_ati_080509.pdf, downloaded 3 November 2009.

22.   Reported sightings or signs of exotic species compiled by DEFRA Rural Development Service (2001 to September 2006) and Natural England (from October 2006)", DEFRA, http://docs.google.com/gview a=v&q=cache:wi17rxoY55AJ:www.naturalengland.org.uk/Images/exoticsstatssummary_tcm6-4154.pdf, accessed 10 November 2009

23.   Avon and Somerset Police, 9 March 2007, http://www.avonandsomerset.police.uk/information/foi/QandA_Question.aspx?qid=174, accessed 4 November 2009.

24.   BBC Gloucestershire, http://www.bbc.co.uk/gloucestershire/content/articles/2006/03/01/big_cat_pauntley_feature.shtml, accessed 6 November 2009.

25.   This is Gloucestershire, http://www.thisisgloucestershire.co.uk/latestnews/Big-cat-sighting-A417/article-224848-detail/article.html, accessed 8 November 2009.

26.   Francis, Di., *Cat Country: The Quest for the British Big Cat*, David and Charles, 1983, pp. 116-128.

27.   Mc Kinlay, Megan, "Churchill's Black Dog? The history of the Black Dog as a Metaphor for depression", Black Dog Institute, January 2005, http://www.blackdoginstitute.org.au/docs/McKinlay.pdf, accessed 8 November 2009.

28.   Dickens, Charles, *The Uncommercial Traveller*, Chapman and Hall, 1861, p. 234.

29.   Palmer, Roy, *The Folklore of Worcestershire*, Logaston Press, 2005, p. 158.

30.   Jenkins, Stephen, *The Undiscovered Country, Adventures into other Dimensions*, Neville Spearman, 1977, p. 177.

31.   BBC Gloucestershire, http://www.bbc.co.uk/gloucestershire/content/articles/2006/10/25/black_dogs_feature.shtml, accessed 7 November 2009.

32.   Woodward, James, M., *The History of Bordesley Abbey in the Valley of the Arrow*, J.H. and H. Parker, 1866, p. 95.

33.   Bord, Janet, "Some Fortean Ramblings", *Fortean Times*, 7, 1974, 6-8.

34.   Paranormal Database, http://www.paranormaldatabase.com/gloucestershire/glosdata.php?pageNum_paradata=2&totalRows_paradata=126, accessed 9 November 2009.

35.   Partridge, J. B., "Cotswold Lore and Customs", *Folk-Lore*, 23,340.

36.   Palmer, Roy, *Folklore of Gloucestershire*, Westcountry Books, p. 160.

37.   Farson, Dan, *The Hamlyn Book of Ghosts in Fact and Fiction*, Hamlyn, 1978, p. 31.

38.   Mysterious Britain, http://www.mysteriousbritain.co.uk/england/gloucestershire/hauntings/headless-black-dog.html, accessed 9 November 2009.

39.   The account of the three counties group is available on their website at http://

www.3cpo.org.uk/html/st__briavels_castle.html, accessed 26 November 2009.

40. Palmer, *Folklore of Gloucestershire*, p. 162.
41. Ibid., p. 165.
42. Turner, M., *Folklore and Mysteries of the Cotswolds*, 1993, p. 175-76.
43. Anon, *A Six Days Tour Through the Isle of Man*, William Dillon, 1836, briefly discusses the belief with regard to Peel Castle.
44. Palmer, *Folklore of Gloucestershire*, p. 157.
45. Potter, F. S., "Gloucestershire Legends", *Folklore*, 25, 3, 1914, 374- 375.
46. Allies, Jabez, *The Ancient British Roman and Saxon Antiquities and Folk-Lore of Worcestershire*, J. H. Parker, London, 1852, pp. 442-443.
47. Palmer, *Folklore of Gloucestershire*, p. 42.
48. Guest, Charlotte, translated, *The Mabinogion*, 1849, reprinted Echo Library, 2006, pp. 99-100.
49. Palmer, *Folklore of Gloucestershire*, p. 32.
50. Ibid., p. 41.
51. Potter, "Gloucestershire Legends", 374-75
52. Palmer, *Folklore of Gloucestershire*, p.146.
53. Giles, John Allen, *William of Malmesbury's Chronicle*, Henry G. Bohn, 1847, pp. 230-231.
54. Palmer, *Folklore of Gloucestershire*, p. 91.
55. Apocalypse, 12.7.
56. Barber, Richard, translated, *Bestiary, being an English Version of the Bodleian Library, Oxford, M.S.Bodley 764, with all the original miniatures reproduced in facsimile*, Woodbridge, 1999, pp. 63-64.
57. Berkeley, Grantley Fitzhardinge, *Berkeley Castle, an Historical Romance*, Volume 1, Richard Bentley, 1836, p. 12.
58. Palmer, *Folklore of Gloucestershire*, p. 144.
59. Ibid., p. 145.
60. Fodor, Nandor, *Between Two Worlds*, Parker Publishing Company, 1965, pp. 172-73.
61. Ibid., p. 153.
62. Palmer, *The Folklore of Worcestershire*, p. 189.
63. Allies, *Ancient British Roman and Saxon Antiquities*, p. 115.
64. *Hansard*, 25 February 2004, c65w.
65. Currie, C R J., Herbert, N M, Edited, *A History of the County of Gloucestershire*, Volume 5, Victoria County History 1996, p. 289.
66. BBC News, 9 December 2002, http://news.bbc.co.uk/1/hi/england/2557739.stm, accessed 20 November 2009.
67. British Wild Boar website, http://www.britishwildboar.org.uk/, downloaded 24 April 2009.
68. Reported sightings or signs of exotic species compiled by DEFRA Rural Development Service.
69. Ibid.
70. *Hansard*, 26 April 2004, c75w.
71. *The Independent*, 23 December 2008, p. 12.
72. *The Sun*, 8 August 1997, p. 28.

73. Fort, Charles, *Lo*, Claude Kendall, 1931, p. 79.
74. British Waterways news release, http://www.britishwaterways.co.uk/newsroom/all-press-releases/display/id/1341, 14 November 2005, downloaded 6 May 2009.
75. Reported sightings or signs of exotic species compiled by DEFRA Rural Development Service.
76. Unexplained Mysteries, forums, http://www.unexplained-mysteries.com/forum/index.php?showtopic=115433, accessed 21 November 2009.
77. *The Graphic*, June 11, 1881.
78. Fort, *Lo*, pp. 549-551.
79. Anon, *Strange and true news from Gloucester*, 2 August, 1660.
80. *Animals & Men*, 6, April 1995, 8.
81. Fraser, Mark, edited, *Big Cats in Britain Yearbook 2007*, p. 63
82. Forest, Gordon, Needham, Mervyn and Birks, Johnny, "A 2002 Pine Marten Record for Worcestershire", *Worcestershire Record*, 13, November 2002, p. 23.
83. BBC News, 23 August 2009, http://news.bbc.co.uk/1/hi/england/gloucestershire/8216856.stm, accessed 3 September 2009.
84. Whitehead, P. F., Limax maculates, (Kaleniczenko, 1851) (Mollusca, Limadidae) New to Worcestershire, *Worcestershire Record*, 11, November 2001, p. 40.
85. Green, David M. "New Snail to Worcestershire – Hygromia Cinctella", *Worcestershire Record*, 4, May 1998, p. 5.
86. Partridge, John, "Spiders: Last Year, Recorders' Day and What next?" *Worcestershire Record*, 6, April 1999, 13.
87. *Birmingham Post*, 30 January 2002, p. 6.
88. Shepherd, Alan, "Worcestershire Turtles and Crawling King Snakes", *Worcestershire Record*, 11, 2001, 34-36.
89. Day, John, Worcestershire Terrapins, *Worcestershire Record*, 10, April 2001, p. 8.
90. Reptile and Amphibian Report 2002, by Colin Twissell, Gloucestershire Naturalists Society, http://www.glosnats.org.uk/reports/reptile.php, accessed 5 November 2009.
91. "Gloucestershire Police Team Up with Nigel Marven to Fight the Illegal Trade in Wild Animals", *Cotswold News* website, http://www.cotswolds.org/cotswolds_news.asp, accessed 5 November 2009.
92. Gloucester Paranormal and Fortean Investigations, http://parafort.com/?page_id=173, downloaded 21 April 2009.
93. Peplow, Gavin, "Birds in Worcestershire", *Worcestershire Record*, 15, November 2003, 13.
94. A full report is on the website of the birds of South Gloucestershire, http://www.thebirdsofsouthgloucestershire.co.uk/SG%20Firsts%20Reports/Cory's%20Shearwater.htm, accessed 22 March 2010.
95. This photo can be seen on the *Wildlife Extra* website, http://www.wildlifeextra.com/go/news/egyptian-goose378.html, accessed 18 March 2010.
96. *Wildlife Extra*, http://www.wildlifeextra.com/go/news/slimbridge-glacous.html#cr, accessed 18 March 2010.
97. Castlemorton website, http://castlemorton.malvernhills.org.uk/pages/wildlife.php, downloaded 10 May 2009.
98. Gloucestershire Naturalists Society, http://www.glosnats.org.uk/reports/

ornithological.php, downloaded 29 April 2009.

99.   Stretch, Brian, "Review of the Year 2007, "Worcestershire Birding, http://
       worcesterbirding.co.uk/26.html, downloaded 28 March 2010. Elsewhere on the
       Worcestershire Birding website this is given as Grimeley, 8-9 October 2007.

100.   The Forest's Strange goat-milkers", BBC News, 11 October 2007, http://
       www.bbc.co.uk/gloucestershire/content/articles/2005/06/20/nightjars_feature.shtml,
       accessed 10 September 2009.

101.   Ospreys Project, http://www.ospreys.org.uk/Sightings%20Spring%202000.htm, ac-
       cessed 18 March 2010.

102.   Peplow, Gavin, "Birds in Worcestershire November 2005 to March 2006", *Worcester-
       shire Record*, 20, April 2007, 31.

103.   *The Guardian*, 7 February 2010.

104.   Peplow, Gavin, "Birds in Worcestershire May to October 2007,", *Worcestershire Re-
       cord*, 23, November 2007, 27.

105.   Monk, James., "The past and present status of the Wryneck in the British Isles", *Bird
       Study*, 10.2, 1963, 123.

106.   *Ibid.*, 126.

107.   Westwood, B., "Worcestershire Wolves", *Worcestershire Record*, 3, November 1997,
       7.

108.   Mc Gee, Kevin, "Dolichovespula media (Hymenoptera: Vespidae). Records of the Un-
       usual Social Wasp in Worcestershire", *Worcestershire Record*, 11, November 2001, 31.

109.   Trevis, Geoff, "Bombus Ruderatus in Worcestershire?", *Worcestershire Record*, 6,
       April 1999, 19.

110.   Green, David M., "Lycoperdina Bovistae (Endomychidae), *Worcestershire Record* 3,
       November 1997, 3.

111.   Green, H, and Meiklejohn, J., "Mistletoe Bugs and a Weevil: Ixapion Variegatum in
       Worcestershire", *Worcestershire Record*, 17 November 2004, 24-25.

112.   Meiklejohn, John, Partridge, John and Green, Harry, "Bombardier Beetle Brachinus
       Crepitans found near Honeybourne", *Worcestershire Record*, 11, 2001, 44.

113.   *Wildlife Extra*, http://www.wildlifeextra.com/go/news/capricorn-beetle352.html,
       downloaded 29 April 2009.

114.   Williams, Mike, "Essex Skipper arrives in Worcestershire", *Worcestershire Record*, 3,
       November 1997, 8.

115.   Green, Harry, Long-Tailed Blue Lampides boeticus (Linnaeus) in Pershore, *Worcester-
       shire Record*, 9, 2000, 24.

116.   *Wildlife Extra*, http://www.wildlifeextra.com/go/news/butterfly-adonis.html,
       downloaded 29 April 2009.

117.   Knapp, John Leonard, *The Journal of a Naturalist*, Carey & Lea, 1831, pp. 198-99.

118.   Scott-Langley, David, "A Millipede new to Gloucestershire", Gloucestershire Natural-
       ists Society, http://www.glosnats.org.uk/reports/cyl_vul.php, downloaded 29 April
       2009.

119.   Meiklejohn, John "The Long-Winged Conehead in Worcestershire", *Worcestershire
       Record*, 15 November 2003, 36.

120.   Meiklejohn, John, "An Alien in South Worcestershire," *Worcestershire Record*, 5, No-
       vember 1998, 13.

121. Averill, Mike, "Dragonfly Round Up for 2000", *Worcestershire Record*, April 2001, 18.

122. Twissell, Ingrid, "A New Dragonfly for Gloucestershire", Gloucestershire Naturalists Society website, June 2004, http://www.glosnats.org.uk/reports/lib.php, downloaded 29 April 2009.

123. Iliff, David, "Hoverfly Report 2002", Gloucestershire Naturalists Society website, http://www.glosnats.org.uk/reports/hoverfly.php, accessed 5 November 2009.

124. Green, David M., "Asilus Crabroniformis Hornet Robber Fly", *Worcestershire Record*, 5 November 1998, 10.

125. Fort, *Lo*, p. 288.

126. "Asian Tiger Mosquito", Health Protection Agency, http://www.hpa.org.uk/HPA/Topics/InfectiousDiseases/InfectionsAZ/1191942145083/, accessed 19 December 2009.

127. Simpson, A. N. B., Worcestershire Moth Report 2000, *Worcestershire Record*, 10, April 2001, 19.

128. Homan, R., "The Distribution of two "new" Micro-Moths in Worcestershire, *Worcestershire Record* 20 April 2007, 16-18.

129. Goddard, Don, "An Unusual Water Bug from the River Teme: Aphelocheirus Aestivalis (F), *Worcestershire Record*, 4, May 1998, 11.

BIBLIOGRAPHY

Note to sources

Many of the texts cited are available online. For reasons of space and, because it is impossible to click on a hyperlink within a book, I have referenced the original printed version, except where I only viewed the online version and did not have sufficient information to properly reference the original.

Hansard is available via the House of Commons website or the excellent, They Work For You website. Some newspaper reports have been copied onto the Big Cats in Britain website or the CFZ forums. Most newspapers have online archives going back three years. Earlier periods can usually be obtained by contacting the publication directly or visiting a local library. Some libraries provide access to an online database of newspapers going back to 1991 and/or a separate database covering all of the nineteenth century. There is also the newspaper library at Collindale. The Worcestershire Record publishes abridged versions of their articles on the internet. Both they and the Gloucestershire Naturalists Society produce printed reports on various topics on a regular basis.

Google Books have put the text of some rare books online and other companies have done the same. Whilst this makes life easier for the researcher it does remove some of the thrill of handling old volumes, of peering over faded text and making handwritten notes on its content.

Sources

Allies, Jabez, *The Ancient British Roman and Saxon Antiquities and Folk-Lore of Worcestershire*, J. H. Parker, London, 1852.
Anon, *Strange and True Newes from Gloucester*, 2 August 1660.
Anon, *A Six Days Tour Through the Isle of Man*, William Dillon, 1836.

Big Cats and Britain's Ecology, Notes from the discussion at the British Association of Nature Conservationists' (BANC) workshop held at the Oak Hall, Keynes Country Park, Cotswold Water Park, Glos. on 9th September 2006, http://209.85.229.132/search? q=cache:c9YAkMH3XrYJ:www.banc.org.uk/Events/BigCats/Big%2520Cats%2520%26% 2520Britain%27s%2520Ecology.pdf+%22frank+tunbridge%22+cat&cd=6&hl=en&ct=clnk, accessed 22 April 2009.

Atkyns, Sir Robert, *The Ancient and Present State of Gloucestershire,* 1712, reprinted E. P. Publishing 1974.

Averill, Mike, "Dragonfly Round Up for 2000", *Worcestershire Record,* April 2001, 18.

Barber, Richard, translated, *Bestiary, being an English Version of the Bodleian Library, Oxford, M.S.Bodley 764, with all the original miniatures reproduced in facsimile,* Woodbridge, 1999.

Beart, John, "Every Village Should Have One", in *Big Cats in Britain Yearbook 2008,* edited Mark Fraser, CFZ Press, 2009, pp. 19-22.

Berkeley, Grantley Fitzhardinge, Berkeley Castle, an Historical Romance, Volume 1, Richard Bentley, 1836,

Bord, Janet, "Some Fortean Ramblings", *Fortean Times,* 7, 1974, 6-8.

Bord, Janet and Bord, Colin, *Alien Animals,* Stackpole, 1981.

Boyd, Hugh and Eltringham, Stuart Keith. (1962) 'The Whooper Swan in Great Britain', *Bird Study,* 9.4, 1962, 217-241.

Brewer, Ebenezer Cobham, *Brewer's Dictionary of Phrase and Fable,* revised edition 1894, reprinted Ware, 19893.

Chipperfield, Mary, *Lion Country,* Hodder and Stoughton, 1972.

Cowie, Mrs, "Phantom Coaches in England", *Folklore,* 53, 4, 1942, 215-218.

Currie, C R J., Herbert, N M, Edited, *A History of the County of Gloucestershire,* Volume 5, Victoria County History 1996.

Cusdin, P. A., "The Keeping of Wolf-Hybrids in Great Britain", June 2000, http:// www.defra.gov.uk/wildlife-pets/wildlife/protect/documents/dwa-wolfdogs.pdf, accessed 3 November 2009.

Dickens, Charles, *The Uncommercial Traveller,* Chapman and Hall, 1861.

Eberhart, George, M., *Mysterious Creatures, a guide to Cryptozoology,* ABC-Clio, 2002.

Farson, Dan, *The Hamlyn Book of Ghosts in Fact and Fiction,* Hamlyn, 1978.

Fodor, Nandor, *Between Two Worlds,* Parker Publishing Company, 1965.

Forest, Gordon, Needham, Mervyn and Birks, Johnny, "A 2002 Pine Marten Record for Worcestershire", *Worcestershire Record,* 13, November 2002, 23.

Fort, Charles, *Lo,* 1931, reprinted Cosimo Inc., 2006.

Francis, Di, *Cat Country: The Quest for the British Big Cat,* David and Charles, 1983.

Fraser, Mark, edited, *Big Cats in Britain Yearbook 2006,* CFZ Press, 2006.

Fraser, Mark, edited, *Big Cats in Britain, Yearbook 2007,* CFZ Press, 2007.

Fraser, Mark, edited, *Big Cats in Britain, Yearbook 2008,* CFZ Press, 2008.

Gervase of Tilbury, *Otia Imperialia,* edited and translated by S.E. Banks and J. W. Binns, Oxford Mediaeval Texts, 2002.

Giles, John Allen, *William of Malmesbury's Chronicle,* Henry G. Bohn, 1847.

Goddard, Don, "An Unusual Water Bug from the River Teme: *Aphelocheirus aestivalis* (F)", *Worcestershire Record,* 4, May 1998, 11.

Green, David M. "New Snail to Worcestershire – *Hygromia cinctella*", *Worcestershire Record* 4, May 1998, 5.

Green, David M., "*Lycoperdina bovistae (Endomychidae)*, *Worcestershire Record* 3, November 1997, 3.

Green, David M, "*Asilus crabroniformis* Hornet Robber Fly", *Worcestershire Record*, 5 November 1998, 10.

Green, Harry, "Long-Tailed Blue *Lampides boeticus* (Linnaeus) in Pershore", *Worcestershire Record*, 9, 2000, 24.

Green, H, and Meiklejohn, J., "Mistletoe Bugs and a Weevil: *Ixapion variegatum* in Worcestershire", *Worcestershire Record*, 17 November 2004, 24-25.

Greening, Mervyn, "Gloucestershire Naturalist's Society, Mammal Report, 1997", http://www.glosnats.org.uk/reports/mammal.php, accessed 10 November 2009.

Guest, Charlotte, translated, *The Mabinogion*, 1849, reprinted Echo Library, 2006.

Harbird, Richard, "The Return of the Magnificent Seven", *Worcestershire Record*, 4, May 1998, 12.

Harmer, Chris, "Big Cats in the Horsley Valley", *The Fountain*, 41, Winter 2007, pp. 4-5.

Harrison, Graham and Harrison, Janet, *The New Birds of the West Midlands, Covering Staffordshire, Warwickshire, Worcestershire and the Former West Midlands County*, West Midland Bird Club, 2005.

Harte, Jeremy, "Black Dog Studies", in Trubshaw, Bob, edited, *Explore Phantom Black Dogs*, Heart of Albion, 2005, pp. 5-20.

Harte, Jeremy, "The black dog in England: a bibliography", in Trubshaw, Bob, edited, *Explore Phantom Black Dogs*, Heart of Albion Press, 2005, pp. 98-128.

Hartland, Edwin, *The Science of Fairy Tales*, 1891, reprinted by Book Jungle, 2007.

Harvey Bloom, James, *Folklore, old customs and superstitions in Shakespeare Land*, Mitchell, Angus & Clark, 1930.

Herbert, N. M., *A History of the County of Gloucester*, Volume 5, Victoria County History, 1996.

Homan, R, "The Distribution of two "new" Micro-Moths in Worcestershire", *Worcestershire Record* 20, April 2007 p. 16-18.

Huet, Chris, *The Dark Companion: The Origin of "Black Dog" as a Description for Depression*, Black Dog Institute, January 2005. http://docs.google.com/viewer?a=v&q=cache:32Q3YrAc4AIJ:www.blackdoginstitute.org.au/docs/Huet.pdf+black+dog+fears&hl=en&pid=bl&srcid=ADGEEShfR6z_Os8hTYfcFRqBDNx4uV3mnqpOj5KqnJxe6U2w9L25AdRH1yLeCSMQKR6OPfiW7PmO7INWAl8tvK22k1iTKWUitwIKnGCOgOYruTg3yOhh9iKAKKnc4xupyeQt0xxEUUt&sig=AHIEtbSzaOy0Nx12tkWze1Co3qgTE--MLw, accessed 1 December 2009.

Iliff, David, "Hoverfly Report 2002", Gloucestershire Naturalists Society, http://www.glosnats.org.uk/reports/hoverfly.php, accessed 5 November 2009.

Jenkins, Stephen, *The Undiscovered Country, Adventures into other Dimensions*, Neville Spearman, 1977.

Knapp, John Leonard, *The Journal of a Naturalist*, Carey & Lea, 1831.

Mathews, Marcus, *Big Cats Loose in Britain*, CFZ Press, 2007.

Mc Gee, Kevin, "*Dolichovespula media* (Hymenoptera: Vespidae). Records of the Unusual Social Wasp in Worcestershire", *Worcestershire Record*, 11, November 2001, 31

Mc Kinlay, Megan, *Churchill's Black Dog? The history of the Black Dog as a Metaphor for depression*, Black Dog Institute, January 2005.

Meiklejohn, John, "An Alien in South Worcestershire," *Worcestershire Record*, 5, November 1998, 13.

Meiklejon, John "The Long-Winged Conehead in Worcestershire", *Worcestershire Record*, 15 November 2003, 36.

Meiklejohn, John, Partridge, John and Green, Harry, "Bombardier Beetle *Brachinus crepitans* found near Honeybourne", *Worcestershire Record*, 11, 2001, 44.

Minter, Rick, "Big Cats – So what?" in *Big Cats in Britain Yearbook 2007*, edited Mark Fraser, CFZ Press, 2008, pp. 7-14.

Minter, Rick, "If there were Big Cats, I'd see them", in *Big Cats in Britain Yearbook 2008*, edited Mark Fraser, CFZ Press, 2009, pp. 23-30.

Monk, James, "The past and present status of the Wryneck in the British Isles", *Bird Study*, 10.2, 1963, 112-132.

Ogilvie, Jen, "Alien Big Cat Survey 2003-2006 England", *Fortean Times*, May 2007, http://www.forteantimes.com/strangedays/cryptozoology/458/alien_big_cat_survey_20032006_england.html, accessed 1 August 2009.

Palmer, Roy, *Folklore of Gloucestershire*, Westcountry Books, 2001.

Palmer, Roy, *The Folklore of Worcestershire*, Logaston Press, 2005,

Partridge, J. B., "Cotswold Lore and Customs", *Folk-Lore*, 23, 332-342.

Partridge, John, "Spiders: Last Year, Recorders' Day and What next?" *Worcestershire Record*, 6, April 1999, 13.

Peplow, Gavin, "Birds in Worcestershire", *Worcestershire Record*, 15, November 2003, 13.

Peplow, Gavin, "Birds in Worcestershire November 2005 to March 2006", *Worcestershire Record*, 20, April 2007, 31.

Peplow, Gavin, "Birds in Worcestershire May to October 2007," *Worcestershire Record*, 23, November 2007, 27.

Potter, F. S., "Gloucestershire Legends", *Folklore*, 25, 3, 1914, 374-75.

Rennie, James, *The Field Naturalist Volumes 1-2*, Orr and Smith, 1833

Rogers, Liam, *The Wild Hunt,* Samhain, 1999.

Rogers, Michael J., "Report on Rare Birds in Great Britain in 1977", http://www.bbrc.org.uk/1977report.pdf, accessed 23 April 2010.

Shepherd, Alan, "Worcestershire Turtles and Crawling King Snakes", *Worcester Record*, 11, 2001, 34-36.

Sherwood, Simon, "A psychological approach to Black Dogs", in Trubshaw, Bob, edited, *Explore Phantom Black Dogs*, Heart of Albion Press, 2005, pp. 21-35.

Simpson, Jacqueline and Round, Steve, *A Dictionary of English Folklore*, Oxford University Press, 2000.

Simpson, A. N. B., "Worcestershire Moth Report 2000", *Worcestershire Record*, 10, April 2001, 19.

Smith, Guy Newman, *Hunting Big Cats in Britain*, Black Hill Books, 2000.

Stephens, Andy, "A report on the alleged Crocodilian sighting in the Gloucester and Sharpness Canal", Centre for Fortean Zoology, http://www.cfz.org.uk/expeditions/03croc/gcroc.htm, accessed 30 September 2009.

Stewart, Peter, "The Baltic Gull in Gloucestershire – the first confirmed British record", *Birding World,* 20, 4, 2007, pp. 152-153.

Stone, Alby, "Infernal Watchdogs", in Trubshaw, Bob, edited, *Explore Phantom Black Dogs,* Heart of Albion Press, 2005, pp. 36-56.

Stretch, Brian, "Review of the Year 2007, "Worcestershire Birding, http:// worcesterbirding.co.uk/26.html, accessed 28 March 2010.

Swale, Trystan, "Gloucestershire Big Cats", in *Big Cats in Britain Yearbook 2006,* edited Mark Fraser, CFZ Press, 2007, pp, 27-31.

Swanton, Michael, edited and translated, *The Anglo-Saxon Chronicle,* Routledge, 1998.

Trubshaw, Bob, edited, *Explore Phantom Black Dogs,* Heart of Albion Press, 2005.

Trevis, Geoff, "*Bombus ruderatus* in Worcestershire?", *Worcestershire Record,* 6, April 1999, 19.

Trevis, Geoff, "The times they are a-changing", *Worcestershire Record,* 26 April 2009, pp. 11-13.

Tunbridge, Frank, "Black Panther killed on bypass", in *Big Cats in Britain Yearbook 2008,* edited Mark Fraser, CFZ Press, 2008, pp. 35-40

Turner, Mark, *Folklore and Mysteries of the Cotswolds,* Robert Hale, 1993, p. 175-76.

Twissell, Ingrid, "A New Dragonfly for Gloucestershire", Gloucestershire Naturalists Society website, June 2004, http://www.glosnats.org.uk/reports/lib.php, accessed 29 April 2009.

Westwood, B., "Worcestershire Wolves", *Worcestershire Record,* 3, November 1997

Whitehead, P. F., "*Limax maculates,* (Kaleniczenko, 1851) (Mollusca, Limadidae) New to Worcestershire", *Worcestershire Record,* 11, November 2001, 40.

Woodward, James, M., *The History of Bordesley Abbey in the Valley of the Arrow,* J.H. and H. Parker, 1866.

Williams, Mike, "Essex Skipper arrives in Worcestershire", *Worcestershire Record,* 3, November 1997, 8.

Williams, Jan, "British Big Cats- The Early Years", in *Big Cats in Britain Yearbook 2006,* edited Mark Fraser, CFZ Press, 2007, pp. 7-12.

Winnall, Rosemary, "Otters in Kidderminster", *Worcestershire Record,* 26, April 2008, 8.

Pilot pub and pull in with the author's Landrover (Maisie)

APPENDIX ONE

A report on the alleged Crocodilian sighting in the Gloucester And Sharpness Canal
by Andy Stephens

Background
I first visited the Canal on Thursday 26th June, following an email request from Jon on the 23rd, asking me to investigate a report of a possible crocodilian attack on a waterfowl witnessed by Richard Lacy, a bridgekeeper, at 'Sellers Bridge', and reported in a press cutting from the Gloucester Citizen dated 19th June entitled 'See you later alligator...?'.

Location
The area of cryptozoological interest is centred on 'Sellars Bridge', a manned swing-bridge allowing pedestrian and vehicular access from the south to the north side of the waterway. The bridge is one of several crossing the canal along the stretch immediately south of Gloucester. The next nearest bridge to Gloucester is Rea Bridge, which is followed by Hempsted Bridge, very close to the city centre.

Sellars Bridge can be found by taking the Junction 12 exit from the M5, proceeding onto the southbound A38 (signposted Stroud) and then immediately taking the minor road/lane on the right hand side signposted 'Hardwicke Church'. Bear right around the Church corner following the lane until after approx ¼ mile or so you reach a mini-roundabout, where you turn left. Sellars Bridge is a few hundred yards down this road past the Pilot public house. The Pilot, situated on the south bank (see photo), overlooks the Bridge and the specific stretch of water of interest in this case. There is a small gravelled pull-in on the left just past the pub where 3 or 4 vehicles can park up.

Opposite the pub on the northern bank there is a small bridgeman's hut accessible by crossing the Sellars Bridge on foot or by car. An attendant can normally be found during normal working, daylight hours, either in this hut or on the swingbridge itself.

Important note for potential researchers
The British Waterways Authority, as owners of the Canal, and the employers of the bridge attendants, have instructed them, and indeed any prospective interviewers of them, to gain permission prior to any interviews taking place.

Bridgemans hut and canal at Sellars

This is entirely legitimate with them acting as a responsible organisation to ensure the effective management of any public safety issues that these reports may contain.

In view of this, I would respectfully request that all prospective investigators cooperate fully with the Authority's request by seeking permission, and by proactively disclosing any significant findings from their research that would assist the authority in maintaining public safety.

In this regard, I would like to hereby acknowledge the assistance given by the Authority in agreeing to my interview with Richard Lacy.

First Visit Thursday 26th June
I arrived at approx 1.30pm, on a brilliant sunny day with temperatures in the 80s.

I first met Eric Perkins, 69, a retired bridgekeeper for more than 26 years, who was tending the garden of Bridgekeepers Cottage, which overlooks Sellars Bridge. He was initially defensive until I explained a little about the role of the Centre and my wish to merely collect the facts as they stood. Clearly temperatures were running a little hot, as the area had been visited by hordes of reporters looking to poke fun at the whole thing and, on one occasion, by a TV film crew who had turned up with their own green, plastic croc in tow!

Eric said he had been aware of reports about something unusual in the water for some time, but he hadn't seen anything himself. He was personally sceptical but acknowledged that the key witness was very experienced on the waterway, and certainly had a good knowledge of the local fish, including obvious candidates such as pike.

On the general topic of out of place animals Eric said there were terrapins living further down the canal, and quite recently he had seen an attack on a rabbit by a 'polecat', which is interesting as this is the second report of polecats I have had from the area. However, I was pursuing the 'croc-o-story' as Jon called it, so I sought out the bridgekeeper's hut where I discovered that, unfortunately, Richard was not on duty that day.

Rather than waste the visit I decided to take my camera and examine the animal life on and in the canal immediately around Sellers bridge (see below).

Second Visit, 7th July, including an interview with Richard Lacy

On my second visit I was told by the bridgekeeper of the day that Richard was to be found on duty at Rea, so I drove the couple of miles up to this location, where Richard was already aware of my visit. He had spoken to the Authority and already obtained agreement to giving an interview, in return for which I complied with the request to leave details of who I was, the organisation I represented, and the reason for our interest, along with my telephone number.

I had a preliminary general chat with Richard who said that his fleeting observation had been made through the window of the bridgekeeper's hut at Sellars. He was keen for me to understand that he had extensive knowledge of the waterway and its fish life, and in particular that what he had seen was a 'scaly' animal, more or less completely airborne, and with legs ' it was no pike'.

I reiterated my position that I was not seeking to discredit him, or his observations, merely to record them and produce a report for the Centre.
He explained that following the sighting it was suggested to him that he should contact the press, which he did, an action which he obviously regrets, as he had little idea of the media circus which was to follow.

We followed the introductions with some further general chat about the physical characteristics of the canal, and its wildlife, of which more below. With these preliminaries over, I switched on the tape recorder. What follows is a verbatim record. Any explanatory notes have been added in italics.

R: So shall I start from the beginning then?
A: So your name is Richard Lacy, you live on the canal here?
R: I live on a boat on the canal down at Slimbridge (some miles west of Sellars)
A: Right, ok, and your age is?
R: 56, and I've been a bridgekeeper for years, I've been on the canal as a boat owner for 40 years; I've also been a coarse fisherman since the age of 8. As far as this sighting was concerned it was 5 to 10 on the morning, sorry, I can't remember what the date was, its gone out of my head, (we can conclude it was shortly before June 23rd when the report appeared in the Citizen) and I was looking southwards down the canal from Sellars, and I was on the phone (in the bridgeman's hut) and there was a very strange disturbance in the water, not a fish-type disturbance, almost a boiling. Conditions were absolutely flat calm, no wind, no sun, very quiet, nobody about. On the edge of the disturbance was a duck, I think it was a mallard, you know a mallard?
A: Yep
R: and, she suddenly…she was acting very strangely, in obvious alarm, then she did a… best described as an emergency takeoff, you know, it wasn't a normal duck takeoff, it was 'Ah, I'm out of here!' (joint laughter)
R: As she went up, um, this… I think its a caiman, but came right up out of the water, at an angle of 45 degrees, it was almost up on it's tail, making a grab for it, and then it just went back down in the water. And it had a sh…it was about 3 foot long, um, I only had sort of 1 1/2 to 2 second view of it, about 3 foot

long, ah, a head from about between about 10 to 12 inches, a short body, and a long tapering tail, and I could see it had two legs on it.

A: And as far as you were concerned the body was split up into 3 definite portions, a head, a body, and a tail?

(interview stopped for the bridgeman's phone ringing)

A: (recommencing interview) So as far as you were concerned it had the three definite portions?
R: Yes
A: And how many legs did you say?
R: Well, I only see the two on my side
A: Ok, so your sort of saw it on a long low angle?
R: It was at an angle of about 45 degrees I would think, that order. It was going away...
A: From one side?
R: From one side, yea.
A: How close did it get to grabbing the bird?
R: 8, 9 inches
A: So it was really quite close?
R: It was ... it was

(Richard gesticulates with his hands - a very short distance)

A: Let me just a...

(interview tape stopped, for Andy to take stock)

A: You were telling me that um, this isn't the first report on something strange in that particular area of the canal.
R: Oh no, that's right, about, I think for the last 4 years we've had odd reports about um, strange movements in the water, um, we've had a couple of people actually say they've seen a caiman or an alligator type-thing in the water. And we've put some of it down to the fact that we've got terrapins down there. They've been in there about 4 years, and they went in round about 2 1/2 to 3 inches across and we've got a couple down there nearly the size of dinner plates now.
A: Yes, last time I was here, actually, which was a week ago, well, Friday the week before last (it was Thursday) I actually wandered down the river because one of the other chaps said you had terrapins and I actually photographed a terrapin on the other, side so I know the terrapins are there.
R: Yes, we've got another set up on the other side of this bridge as well now (northwards from Rea bridge) so they're surviving quite happily.
A: Yes, they're out of place of course they shouldn't be there should they?
R: That's right
A: They, they're not sort of indigenous to the UK
R: We've got all sorts of things out of place around here. We've got little egrets flying around down on the southern end of the canal, I mean, not exactly a local bird is it? (laughter)
A: No absolutely!
(I tried to obtain more specific physical information about the stretch in question)
R: The whole stretch is 15 1/2 miles from Gloucester to Sharpness. Its an average of 14 feet deep, its an average of about 100 feet wide. And the particular cutting we're talking about is only accessible from one side. Its fairly well...um...a large amount of vegetation. There's a lot of bank slips along there which creates a shelf varying from a couple of inches out to just under 2 feet for about 20 feet out into the whole canal, on that, on that length for just under a mile. And there's a number of small culverts, um,

channel culverts, off that section as well. That's about all I can say. on that.
At that point I stopped the tape.

Richard at Rea Bridge

I returned to trying to establish the precise area where the actual sighting took place, which Richard confirmed was on the southern bank, 50 yards or so westerly from the Pilot, more or less opposite the navigation light, where there is a landslip.

He said it was interesting that this stretch was unusual in that it was the only stretch accessible from a single bank, the other being overgrown, and with landslips creating a shelf, and with culverts. He also described it as 'lonely', people do not come here in great numbers – a few walkers on the towpath, a few runners, but there is passing traffic on the water.

I had already observed on my previous visit that the passing of a large passenger vessel had little effect on the fish in the margins, and did not upset the basking terrapins (see below), but when still 100 yards away, had induced a pike (see below), to sink away, slowly into the depths.

Richard said that this report was unusual, as most incidents have been reported later in the evening, whereas this reported attack was in the morning.

Notes on physical characteristics of the canal
The Gloucester and Sharpness Canal is a 151/2 mile long canal starting a mile north of the city of Gloucester and running south and west until it terminates at Sharpness docks, on the banks of the river Severn, some 1 3/4 miles north of the nuclear power station at Berkeley. The Canal is on average 100 feet wide, 14 feet deep in the middle, tapering to 3 or 4 feet in the margins. It is a regular summertime haunt for pleasure boats, long boats, sports rowing boats, anglers and the walkers who take advantage of its extensive towpaths.

A pike just under the water. Could this be mistaken for the forquarters of a crocodilian>
On the water, healthy populations of swans, mallard and coot were going about their business.
In the air large numbers of pigeons were flying between the numerous trees, and swallows
skimmed over the water's surface.

Notes on the canal ecosystem

On walking the stretch on my first visit the first thing I noticed, as a fisherman and amateur naturalist of some 35 years experience, was just how healthy the ecosystem is on this particular stretch of Canal.

Running South and West from the Pilot this stretch of the Canal is, as Richard says, unusual, in that for a mile or so the south bank is overgrown and cannot be reached on foot. The accessible northern bank however has a well-defined towpath/footpath.

Even so, the area between the towpath and the water is generally overgrown with grasses, trees and bushes. In most places the bank is undercut and any attempt to reach the water would be to risk falling in, but there are several places where concrete blocks and/or corrugated steel sheeting have been put in place to reinforce the banking. These allow access to the water's edge.

Standing at these points you can look into the water and see more or less to the bottom. In every case these margins were full of large shoals of fish fry. Looking to mid water small to medium sized perch were visible amongst the weed growth, which is quite light. Below this shoals of mixed size roach and rudd were seen grubbing around in the mud. The fish were as prolific as I have ever seen on any stretch of UK canal.

As I walked westerly and the Pilot pub disappeared from sight round a slight bend I took particular notice of the opposite (South) bank where Eric had advised the colony of terrapins were be found basking on partly submerged logs on hot days. Sure enough I saw a red eared terrapin some 10 inches or so in length basking on the logs, and a photo taken with a Nikon Coolpix 5700 on full zoom accompanies this report.

On returning back northwards and eastwards towards the bridge I saw a medium sized pike about 3 feet long hanging in the water below overhanging branches (see photo). There are obviously much bigger sized pike to be found, Richard Lacy thinks they might grow to 30 pounds or so on this waterway.

The notorious out of place invader, the zander, has unfortunately found its way into the canal, and Richard thinks they may get to the 14lb mark.

Richard tells me that many other species of fish and eels also thrive in the waterway. The banks are the home to the usual small indigenous species such as vole, mice, rats, stoat, weasel, and rabbit, as well as an increasing population of that voracious hunter and invader, the mink.

Final thoughts

It goes without saying that pike in the middle and upper, and zander towards the upper levels of their size band could attack the wildfowl on this stretch. However, I have to acknowledge here the experience that Richard has of the wildlife on the canal. He remains convinced that what he saw was not a pike; indeed his observation of three definite parts to the body, the last one being a long tapering tail are not characteristic of UK freshwater predatory fish.

We should also note how he first saw an odd disturbance in the water, and then, with his extensive knowledge of behaviour, gained from years of wildlife observation at close quarters (he lives on the water for goodness sake!) he was drawn to the unusual behaviour of the duck.

I am worried though, by aspects of the testimony. Richard, with all his experience, was not certain what type of duck it was ('I think it was a mallard'), yet he was convinced what he saw was a caiman-like animal and not a pike. There is an inconsistency here I am uncomfortable with, especially when the observation was snatched, through a window, whilst the observer was answering the phone.

I have absolutely no doubt Richard believes he saw a caiman-like animal, I do not doubt his honesty for a second, and I would want to go on record to remove from the equation in this case, one of the potential causes for some crypto reports, that of dishonesty by a misguided individual.

Playing Devil's advocate, from my short walk, I would have no hesitation in concluding that the ecosystem supports more than adequate numbers of food species to keep a single crocodilian, or indeed a small colony, well fed, as long as they could survive the rigours of the English winter. Clearly the terrapins manage; demonstrably, they have been here for 4 or 5 years.

It is worth recording that I have lived in Cheltenham for 61/2 years (only 10 or so miles from Hardwicke) and all the recent winters I have experienced have been uncharacteristically mild and wet, with few long periods of sub-zero temperatures.

Jon and I agree that no crocodilian could, in all likelihood, thrive in a UK climate without there being a source of heat to maintain its body temperature throughout the winter months. We need to ask along the crucial stretch, what do the culverts drain, could they be the source of any flow into the river during winter, which might raise water temperature?

In the absence of any further information the file must remain open.

STILL ON THE TRACK OF UNKNOWN ANIMALS

The Centre for Fortean Zoology, or CFZ, is a non profit-making organisation founded in 1992 with the aim of being a clearing house for information, and coordinating research into mystery animals around the world.

We also study out of place animals, rare and aberrant animal behaviour, and Zooform Phenomena; little-understood "things" that appear to be animals, but which are in fact nothing of the sort, and not even alive (at least in the way we understand the term).

Not only are we the biggest organisation of our type in the world, but - or so we like to think - we are the best. We are certainly the only truly global cryptozoological research organisation, and we carry out our investigations using a strictly scientific set of guidelines. We are expanding all the time and looking to recruit new members to help us in our research into mysterious animals and strange creatures across the globe.

Why should you join us? Because, if you are genuinely interested in trying to solve the last great mysteries of Mother Nature, there is nobody better than us with whom to do it.

Members get a four-issue subscription to our journal *Animals & Men*. Each issue contains nearly 100 pages packed with news, articles, letters, research papers, field reports, and even a gossip column! The magazine is Royal Octavo in format with a full colour cover. You also have access to one of the world's largest collections of resource material dealing with cryptozoology and allied disciplines, and people from the CFZ membership regularly take part in fieldwork and expeditions around the world.

The CFZ is managed by a three-man board of trustees, with a non-profit making trust registered with HM Government Stamp Office. The board of trustees is supported by a Permanent Directorate of full and part-time staff, and advised by a Consultancy Board of specialists - many of whom are world-renowned experts in their particular field. We have regional representatives across the UK, the USA, and many other parts of the world, and are affiliated with other organisations whose aims and protocols mirror our own.

You'll find that the people at the CFZ are friendly and approachable. We have a thriving forum on the website which is the hub of an ever-growing electronic community. You will soon find your feet. Many members of the CFZ Permanent Directorate started off as ordinary members, and now work full-time chasing monsters around the world.

Write to us, e-mail us, or telephone us. The list of future projects on the website is not exhaustive. If you have a good idea for an investigation, please tell us. We may well be able to help.

We are always looking for volunteers to join us. If you see a project that interests you, do not hesitate to get in touch with us. Under certain circumstances we can help provide funding for your trip. If you look on the future projects section of the website, you can see some of the projects that we have pencilled in for the next few years.

In 2003 and 2004 we sent three-man expeditions to Sumatra looking for Orang-Pendek - a semi-legendary bipedal ape. The same three went to Mongolia in 2005. All three members started off merely subscribers to the CFZ magazine. Next time it could be you!

We have no magic sources of income. All our funds come from donations, membership fees, and sales of our publications and merchandise. We are always looking for corporate sponsorship, and other sources of revenue. If you have any ideas for fund-raising please let us know.

However, unlike other cryptozoological organisations in the past, we do not live in an intellectual ivory tower. We are not afraid to get our hands dirty, and furthermore we are not one of those organisations where the membership have to raise money so that a privileged few can go on expensive foreign trips. Our research teams, both in the UK and abroad, consist of a mixture of experienced and inexperienced personnel. We are truly a community, and work on the premise that the benefits of CFZ membership are open to all.

Reports of our investigations are published on our website as soon as they are available. Preliminary reports are posted within days of the project finishing.

Each year we publish a 200 page yearbook containing research papers and expedition reports too long to be printed in the journal. We freely circulate our information to anybody who asks for it.

We have a thriving YouTube channel, CFZtv, which has well over two hundred self-made documentaries, lecture appearances, and episodes of our monthly webTV show. We have a daily online magazine, which has over a million hits each year.

Each year since 2000 we have held our annual convention - the Weird Weekend. It is three days of lectures, workshops, and excursions. But most importantly it is a chance for members of the CFZ to meet each other, and to talk with the members of the permanent directorate in a relaxed and informal setting and preferably with a pint of beer in one hand. Since 2006 - the Weird Weekend has been bigger and better and held on the third weekend in August in the idyllic rural location of Woolsery in North Devon.

Since relocating to North Devon in 2005 we have become ever more closely involved with other community organisations, and we hope that this trend will continue. We have also worked closely with Police Forces across the UK as consultants for animal mutilation cases, and we intend to forge closer links with the coastguard and other community services. We want to work closely with those who regularly travel into the Bristol Channel, so that if the recent trend of exotic animal visitors to our coastal waters continues, we can be out there as soon as possible.

Apart from having been the only Fortean Zoological organisation in the world to have consistently published material on all aspects of the subject for over a decade, we have achieved the following concrete results:

• Disproved the myth relating to the headless so-called sea-serpent carcass of Durgan beach in Cornwall 1975
• Disproved the story of the 1988 puma skull of

Lustleigh Cleave

- Carried out the only in-depth research ever into the mythos of the Cornish Owlman.
- Made the first records of a tropical species of lamprey
- Made the first records of a luminous cave gnat larva in Thailand
- Discovered a possible new species of British mammal - the beech marten
- In 1994-6 carried out the first archival fortean zoological survey of Hong Kong
- In the year 2000, CFZ theories were confirmed when a new species of lizard was added to the British List
- Identified the monster of Martin Mere in Lancashire as a giant wels catfish
- Expanded the known range of Armitage's skink in the Gambia by 80%
- Obtained photographic evidence of the remains of Europe's largest known pike
- Carried out the first ever in-depth study of the ninki-nanka
- Carried out the first attempt to breed Puerto Rican cave snails in captivity
- Were the first European explorers to visit the `lost valley` in Sumatra
- Published the first ever evidence for a new tribe of pygmies in Guyana
- Published the first evidence for a new species of caiman in Guyana
- Filmed unknown creatures

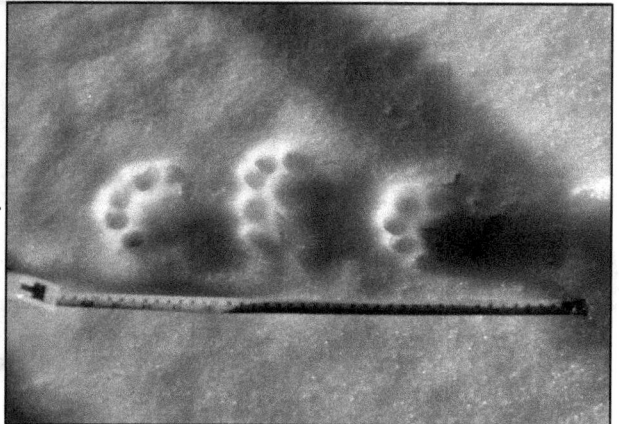

on a monster-haunted lake in Ireland for the first time
- Had a sighting of orang pendek in Sumatra in 2009
- Found leopard hair, subsequently identified by DNA analysis, from rural North Devon in 2010
- Brought back hairs which appear to be from an unknown primate in Sumatra
- Published some of the best evidence ever for the almasty in southern Russia

CFZ Expeditions and Investigations include:

- 1998 Puerto Rico, Florida, Mexico (Chupacabras)
- 1999 Nevada (Bigfoot)
- 2000 Thailand (Naga)
- 2002 Martin Mere (Giant catfish)
- 2002 Cleveland (Wallaby mutilation)

- 2003 Bolam Lake (BHM Reports)
- 2003 Sumatra (Orang Pendek)
- 2003 Texas (Bigfoot; giant snapping turtles)
- 2004 Sumatra (Orang Pendek; cigau, a sabre-toothed cat)
- 2004 Illinois (Black panthers; cicada swarm)
- 2004 Texas (Mystery blue dog)
- Loch Morar (Monster)
- 2004 Puerto Rico (Chupacabras; carnivorous cave snails)
- 2005 Belize (Affiliate expedition for hairy dwarfs)
- 2005 Loch Ness (Monster)
- 2005 Mongolia (Allghoi Khorkhoi aka Mongolian death worm)

- 2006 Gambia (Gambo - Gambian sea monster , Ninki Nanka and Armitage's skink
- 2006 Llangorse Lake (Giant pike, giant eels)
- 2006 Windermere (Giant eels)
- 2007 Coniston Water (Giant eels)
- 2007 Guyana (Giant anaconda, didi, water tiger)
- 2008 Russia (Almasty)
- 2009 Sumatra (Orang pendek)
- 2009 Republic of Ireland (Lake Monster)
- 2010 Texas (Blue Dogs)
- 2010 India (Mande Burung)

For details of current membership fees, current expeditions and investigations, and voluntary posts within the CFZ that need your help, please do not hesitate to contact us.

The Centre for Fortean Zoology,
Myrtle Cottage,
Woolfardisworthy,
Bideford, North Devon
EX39 5QR

Telephone 01237 431413
Fax+44 (0)7006-074-925
eMail info@cfz.org.uk

Websites:

www.cfz.org.uk
www.weirdweekend.org

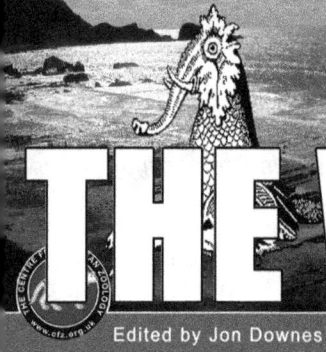

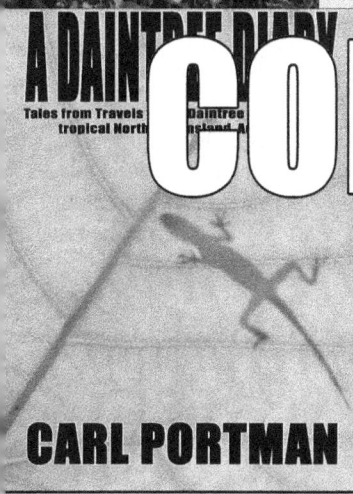
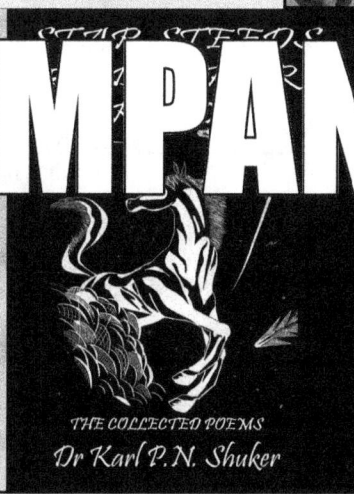

THE WORLD'S WEIRDEST PUBLISHING COMPANY

HOW TO START A PUBLISHING EMPIRE

Unlike most mainstream publishers, we have a non-commercial remit, and our mission statement claims that "we publish books because they deserve to be published, not because we think that we can make money out of them". Our motto is the Latin Tag *Pro bona causa facimus* (we do it for good reason), a slogan taken from a children's book *The Case of the Silver Egg* by the late Desmond Skirrow.

WIKIPEDIA: "The first book published was in 1988. *Take this Brother may it Serve you Well* was a guide to Beatles bootlegs by Jonathan Downes. It sold quite well, but was hampered by very poor production values, being photocopied, and held together by a plastic clip binder. In 1988 A5 clip binders were hard to get hold of, so the publishers took A4 binders and cut them in half with a hacksaw. It now reaches surprisingly high prices second hand.

The production quality improved slightly over the years, and after 1999 all the books produced were ringbound with laminated colour covers. In 2004, however, they signed an agreement with Lightning Source, and all books are now produced perfect bound, with full colour covers."

Until 2010 all our books, the majority of which are/were on the subject of mystery animals and allied disciplines, were published by `CFZ Press`, the publishing arm of the Centre for Fortean Zoology (CFZ), and we urged our readers and followers to draw a discreet veil over the books that we published that were completely off topic to the CFZ.

However, in 2010 we decided that enough was enough and launched a second imprint, `Fortean Words` which aims to cover a wide range of non animal-related esoteric subjects. Other imprints will be launched as and when we feel like it, however the basic ethos of the company remains the same: Our job is to publish books and magazines that we feel are worth publishing, whether or not they are going to sell. Money is, after all - as my dear old Mama once told me - a rather vulgar subject, and she would be rolling in her grave if she thought that her eldest son was somehow in `trade`.

Luckily, so far our tastes have turned out not to be that rarified after all, and we have sold far more books than anyone ever thought that we would, so there is a moral in there somewhere…

Jon Downes,
Woolsery, North Devon
July 2010

CFZ PRESS

Other Books in Print

When Bigfoot Attacks by Michael Newton
Mystery Animals of the British Isles: Gloucester and Worcester by Paul Williams
Weird Waters – The Mystery Animals of Scandinavia: Lake and Sea Monsters by Lars Thomas
The Inhumanoids by Barton Nunnelly
Monstrum! A Wizard's Tale by Tony "Doc" Shiels
CFZ Yearbook 2011 edited by Jonathan Downes
Karl Shuker's Alien Zoo by Shuker, Dr Karl P.N
Tetrapod Zoology Book One by Naish, Dr Darren
The Mystery Animals of Ireland by Gary Cunningham and Ronan Coghlan
Monsters of Texas by Gerhard, Ken
The Great Yokai Encyclopaedia by Freeman, Richard
NEW HORIZONS: Animals & Men *issues 16-20 Collected Editions Vol. 4*
by Downes, Jonathan
A Daintree Diary -
Tales from Travels to the Daintree Rainforest in tropical north Queensland, Australia
by Portman, Carl
Strangely Strange but Oddly Normal by Roberts, Andy
Centre for Fortean Zoology Yearbook 2010 by Downes, Jonathan
Predator Deathmatch by Molloy, Nick
Star Steeds and other Dreams by Shuker, Karl
CHINA: A Yellow Peril? by Muirhead, Richard
Mystery Animals of the British Isles: The Western Isles by Vaudrey, Glen
Giant Snakes - Unravelling the coils of mystery by Newton, Michael
Mystery Animals of the British Isles: Kent by Arnold, Neil
Centre for Fortean Zoology Yearbook 2009 by Downes, Jonathan
CFZ EXPEDITION REPORT: Russia 2008 by Richard Freeman *et al*, Shuker, Karl (fwd)
Dinosaurs and other Prehistoric Animals on Stamps - A Worldwide catalogue
by Shuker, Karl P. N
Dr Shuker's Casebook by Shuker, Karl P.N
The Island of Paradise - chupacabra UFO crash retrievals,
and accelerated evolution on the island of Puerto Rico by Downes, Jonathan
The Mystery Animals of the British Isles: Northumberland and Tyneside by Hallowell, Michael J

Centre for Fortean Zoology Yearbook 1997 by Downes, Jonathan (Ed)
Centre for Fortean Zoology Yearbook 2002 by Downes, Jonathan (Ed)
Centre for Fortean Zoology Yearbook 2000/1 by Downes, Jonathan (Ed)
Centre for Fortean Zoology Yearbook 1998 by Downes, Jonathan (Ed)
Centre for Fortean Zoology Yearbook 2003 by Downes, Jonathan (Ed)
In the wake of Bernard Heuvelmans by Woodley, Michael A
CFZ EXPEDITION REPORT: Guyana 2007 by Richard Freeman *et al*, Shuker, Karl (fwd)
Centre for Fortean Zoology Yearbook 1999 by Downes, Jonathan (Ed)
Big Cats in Britain Yearbook 2008 by Fraser, Mark (Ed)
Centre for Fortean Zoology Yearbook 1996 by Downes, Jonathan (Ed)
THE CALL OF THE WILD - Animals & Men issues 11-15
Collected Editions Vol. 3 by Downes, Jonathan (ed)
Ethna's Journal by Downes, C N
Centre for Fortean Zoology Yearbook 2008 by Downes, J (Ed)
DARK DORSET -Calendar Custome by Newland, Robert J
Extraordinary Animals Revisited by Shuker, Karl
MAN-MONKEY - In Search of the British Bigfoot by Redfern, Nick
Dark Dorset Tales of Mystery, Wonder and Terror by Newland, Robert J and Mark North
Big Cats Loose in Britain by Matthews, Marcus
MONSTER! - The A-Z of Zooform Phenomena by Arnold, Neil
The Centre for Fortean Zoology 2004 Yearbook by Downes, Jonathan (Ed)
The Centre for Fortean Zoology 2007 Yearbook by Downes, Jonathan (Ed)
CAT FLAPS! Northern Mystery Cats by Roberts, Andy
Big Cats in Britain Yearbook 2007 by Fraser, Mark (Ed)
BIG BIRD! - Modern sightings of Flying Monsters by Gerhard, Ken
THE NUMBER OF THE BEAST - Animals & Men issues 6-10
Collected Editions Vol. 1 by Downes, Jonathan (Ed)
IN THE BEGINNING - Animals & Men issues 1-5 Collected Editions Vol. 1 by Downes, Jonathan
STRENGTH THROUGH KOI - They saved Hitler's Koi and other stories by Downes, Jonathan
The Smaller Mystery Carnivores of the Westcountry by Downes, Jonathan
CFZ EXPEDITION REPORT: Gambia 2006 by Richard Freeman *et al*, Shuker, Karl (fwd)
The Owlman and Others by Jonathan Downes
The Blackdown Mystery by Downes, Jonathan
Big Cats in Britain Yearbook 2006 by Fraser, Mark (Ed)
Fragrant Harbours - Distant Rivers by Downes, John T
Only Fools and Goatsuckers by Downes, Jonathan
Monster of the Mere by Jonathan Downes
Dragons:More than a Myth by Freeman, Richard Alan
Granfer's Bible Stories by Downes, John Tweddell
Monster Hunter by Downes, Jonathan

Fortean Words

The Centre for Fortean Zoology has for several years led the field in Fortean publishing. CFZ Press is the only publishing company specialising in books on monsters and mystery animals. CFZ Press has published more books on this subject than any other company in history and has attracted such well known authors as Andy Roberts, Nick Redfern, Michael Newton, Dr Karl Shuker, Neil Arnold, Dr Darren Naish, Jon Downes, Ken Gerhard and Richard Freeman.

Now CFZ Press are launching a new imprint. Fortean Words is a new line of books dealing with Fortean subjects other than cryptozoology, which is - after all - the subject the CFZ are best known for. Fortean Words is being launched with a spectacular multi-volume series called *Haunted Skies* which covers British UFO sightings between 1940 and 2010. Former policeman John Hanson and his long-suffering partner Dawn Holloway have compiled a peerless library of sighting reports, many that have not been made public before.

Other books include a look at the Berwyn Mountains UFO case by renowned Fortean Andy Roberts and a series of forthcoming books by transatlantic researcher Nick Redfern. CFZ Press are dedicated to maintaining the fine quality of their works with Fortean Words. New authors tackling new subjects will always be encouraged, and we hope that our books will continue to be as ground-breaking and popular as ever.

Haunted Skies Volume One 1940-1959 by John Hanson and Dawn Holloway
Haunted Skies Volume Two 1960-1965 by John Hanson and Dawn Holloway
Space Girl Dead on Spaghetti Junction - an anthology by Nick Redfern
I Fort the Lore - an anthology by Paul Screeton
UFO Down - the Berwyn Mountains UFO Crash by Andy Roberts

www.ingramcontent.com/pod-product-compliance
Lightning Source LLC
Chambersburg PA
CBHW052213270326
41931CB00011B/2330